Lecture Notes in Computer Science

Lecture Notes in Computer Science

Vol. 1: GI-Gesellschaft für Informatik e.V. 3. Jahrestagung, Hamburg, 8.–10. Oktober 1973. Herausgegeben im Auftrag der Gesellschaft für Informatik von W. Brauer. XI, 508 Seiten. 1973.

Vol. 2: GI-Gesellschaft für Informatik e.V. 1. Fachtagung über Automatentheorie und Formale Sprachen, Bonn, 9.–12. Juli 1973. Herausgegeben im Auftrag der Gesellschaft für Informatik von K.-H. Böhling und K. Indermark. VII, 322 Seiten. 1973.

Vol. 3: 5th Conference on Optimization Techniques, Part I. (Series: I.F.I.P. TC7 Optimization Conferences.) Edited by R. Conti and A. Ruberti. XIII, 565 pages. 1973.

Vol. 4: 5th Conference on Optimization Techniques, Part II. (Series: I.F.I.P. TC7 Optimization Conferences.) Edited by R. Conti and A. Ruberti. XIII, 389 pages. 1973.

Vol. 5: International Symposium on Theoretical Programming. Edited by A. Ershov and V. A. Nepomniaschy. VI, 407 pages. 1974.

Vol. 6: B. T. Smith, J. M. Boyle, J. J. Dongarra, B. S. Garbow, Y. Ikebe, V. C. Klema, and C. B. Moler, Matrix Eigensystem Routines – EISPACK Guide. XI, 551 pages. 2nd Edition 1974. 1976.

Vol. 7: 3. Fachtagung über Programmiersprachen, Kiel, 5.–7. März 1974. Herausgegeben von B. Schlender und W. Frielinghaus. VI, 225 Seiten. 1974.

Vol. 8: GI-NTG Fachtagung über Struktur und Betrieb von Rechensystemen, Braunschweig, 20.–22. März 1974. Herausgegeben im Auftrag der GI und der NTG von H.-O. Leilich. VI, 340 Seiten. 1974.

Vol. 9: GI-BIFOA Internationale Fachtagung: Informationszentren in Wirtschaft und Verwaltung. Köln, 17./18. Sept. 1973. Herausgegeben im Auftrag der GI und dem BIFOA von P. Schmitz. VI, 259 Seiten. 1974.

Vol. 10: Computing Methods in Applied Sciences and Engineering, Part 1. International Symposium, Versailles, December 17–21, 1973. Edited by R. Glowinski and J. L. Lions. X, 497 pages. 1974.

Vol. 11: Computing Methods in Applied Sciences and Engineering, Part 2. International Symposium, Versailles, December 17–21, 1973. Edited by R. Glowinski and J. L. Lions. X, 434 pages. 1974.

Vol. 12: GFK-GI-GMR Fachtagung Prozessrechner 1974. Karlsruhe, 10.–11. Juni 1974. Herausgegeben von G. Krüger und R. Friehmelt. XI, 620 Seiten. 1974.

Vol. 13: Rechnerstrukturen und Betriebsprogrammierung, Erlangen, 1970. (GI-Gesellschaft für Informatik e.V.) Herausgegeben von W. Händler und P. P. Spies. VII, 333 Seiten. 1974.

Vol. 14: Automata, Languages and Programming – 2nd Colloquium, University of Saarbrücken, July 29–August 2, 1974. Edited by J. Loeckx. VIII, 611 pages. 1974.

Vol. 15: L Systems. Edited by A. Salomaa and G. Rozenberg. VI, 338 pages. 1974.

Vol. 16: Operating Systems, International Symposium, Rocquencourt 1974. Edited by E. Gelenbe and C. Kaiser. VIII, 310 pages. 1974.

Vol. 17: Rechner-Gestützter Unterricht RGU '74, Fachtagung, Hamburg, 12.–14. August 1974, ACU-Arbeitskreis Computer-Unterstützter Unterricht. Herausgegeben im Auftrag der GI von K. Brunnstein, K. Haefner und W. Händler. X, 417 Seiten. 1974.

Vol. 18: K. Jensen and N. E. Wirth, PASCAL – User Manual and Report. VII, 170 pages. Corrected Reprint of the 2nd Edition 1976.

Vol. 19: Programming Symposium. Proceedings 1974. V, 425 pages. 1974.

Vol. 20: J. Engelfriet, Simple Program Schemes and Formal Languages. VII, 254 pages. 1974.

Vol. 21: Compiler Construction, An Advanced Course. Edited by F. L. Bauer and J. Eickel. XIV. 621 pages. 1974.

Vol. 22: Formal Aspects of Cognitive Processes. Proceedings 1972. Edited by T. Storer and D. Winter. V, 214 pages. 1975.

Vol. 23: Programming Methodology. 4th Informatik Symposium, IBM Germany Wildbad, September 25–27, 1974. Edited by C. E. Hackl. VI, 501 pages. 1975.

Vol. 24: Parallel Processing. Proceedings 1974. Edited by T. Feng. VI, 433 pages. 1975.

Vol. 25: Category Theory Applied to Computation and Control. Proceedings 1974. Edited by E. G. Manes. X, 245 pages. 1975.

Vol. 26: GI-4. Jahrestagung, Berlin, 9.–12. Oktober 1974. Herausgegeben im Auftrag der GI von D. Siefkes. IX, 748 Seiten. 1975.

Vol. 27: Optimization Techniques. IFIP Technical Conference. Novosibirsk, July 1–7, 1974. (Series: I.F.I.P. TC7 Optimization Conferences.) Edited by G. I. Marchuk. VIII, 507 pages. 1975.

Vol. 28: Mathematical Foundations of Computer Science. 3rd Symposium at Jadwisin near Warsaw, June 17–22, 1974. Edited by A. Blikle. VII, 484 pages. 1975.

Vol. 29: Interval Mathematics. Procedings 1975. Edited by K. Nickel. VI, 331 pages. 1975.

Vol. 30: Software Engineering. An Advanced Course. Edited by F. L. Bauer. (Formerly published 1973 as Lecture Notes in Economics and Mathematical Systems, Vol. 81) XII, 545 pages. 1975.

Vol. 31: S. H. Fuller, Analysis of Drum and Disk Storage Units. IX, 283 pages. 1975.

Vol. 32: Mathematical Foundations of Computer Science 1975. Proceedings 1975. Edited by J. Bečvář. X, 476 pages. 1975.

Vol. 33: Automata Theory and Formal Languages, Kaiserslautern, May 20–23, 1975. Edited by H. Brakhage on behalf of GI. VIII, 292 Seiten. 1975.

Vol. 34: GI – 5. Jahrestagung, Dortmund 8.–10. Oktober 1975. Herausgegeben im Auftrag der GI von J. Mühlbacher. X, 755 Seiten. 1975.

Vol. 35: W. Everling, Exercises in Computer Systems Analysis. (Formerly published 1972 as Lecture Notes in Economics and Mathematical Systems, Vol. 65) VIII, 184 pages. 1975.

Vol. 36: S. A. Greibach, Theory of Program Structures: Schemes, Semantics, Verification. XV, 364 pages. 1975.

Vol. 37: C. Böhm, λ-Calculus and Computer Science Theory. Proceedings 1975. XII, 370 pages. 1975.

Vol. 38: P. Branquart, J.-P. Cardinael, J. Lewi, J.-P. Delescaille, M. Vanbegin. An Optimized Translation Process and Its Application to ALGOL 68. IX, 334 pages. 1976.

Vol. 39: Data Base Systems. Proceedings 1975. Edited by H. Hasselmeier and W. Spruth. VI, 386 pages. 1976.

Vol. 40: Optimization Techniques. Modeling and Optimization in the Service of Man. Part 1. Proceedings 1975. Edited by J. Cea. XIV, 854 pages. 1976.

Vol. 41: Optimization Techniques. Modeling and Optimization in the Service of Man. Part 2. Proceedings 1975. Edited by J. Cea. XIII, 852 pages. 1976.

Vol. 42: James E. Donahue, Complementary Definitions of Programming Language Semantics. VII, 172 pages. 1976.

Vol. 43: E. Specker und V. Strassen, Komplexität von Entscheidungsproblemen. Ein Seminar. V, 217 Seiten. 1976.

Vol. 44: ECI Conference 1976. Proceedings 1976. Edited by K. Samelson. VIII, 322 pages. 1976.

Vol. 45: Mathematical Foundations of Computer Science 1976. Proceedings 1976. Edited by A. Mazurkiewicz. XI, 601 pages. 1976.

Vol. 46: Language Hierarchies and Interfaces. Edited by F. L. Bauer and K. Samelson. X, 428 pages. 1976.

Vol. 47: Methods of Algorithmic Language Implementation. Edited by A. Ershov and C. H. A. Koster. VIII, 351 pages. 1977.

Vol. 48: Theoretical Computer Science, Darmstadt, March 1977. Edited by H. Tzschach, H. Waldschmidt and H. K.-G. Walter on behalf of GI. VIII, 418 pages. 1977.

Lecture Notes in Computer Science

Edited by G. Goos and J. Hartmanis

101

André Thayse

Boolean Calculus of Differences

Springer-Verlag
Berlin Heidelberg New York 1981

Author

André Thayse
Philips Research Laboratory
av. Van Becelaere, 2
B-1170 Brussels
Belgium

AMS Subject Classifications (1979): 94C05, 94C10
CR Subject Classifications (1979): 6.1

ISBN 3-540-10286-8 Springer-Verlag Berlin Heidelberg New York
ISBN 0-387-10286-8 Springer-Verlag New York Heidelberg Berlin

Library of Congress Cataloging in Publication Data. Thayse, André, 1940-.
Boolean calculus of differences. (Lecture notes in computer science; 101).
Bibliography: p. Includes index. 1. Algebra, Boolean. 2. Switching theory. I. Title.
II. Series. QA10.3.T47. 511.3'24. 80-28632

Printed in Germany

Printing and binding: Beltz Offsetdruck, Hemsbach/Bergstr.
2145/3140-543210

FOREWORD

by Sheldon B. Akers

The development of switching circuit theory over the past three decades has mirrored the varying concerns of the logic designers who have had to confront the many problems presented by constantly changing circuit technologies. All too often, yesterday's elegant solution has been rendered obsolete by today's technological breakthrough. It is not surprising, therefore, that the accepted techniques and procedures of present day switching circuit theory too often tend to stand as distinct entities rather than as part of a cohesive whole.

Accordingly, it is a great pleasure to be able to recommend a book which not only provides a much needed historical perspective to these many developments but, even more importantly, does so within the framework of a single comprehensive structure. Starting with the basic definitions of Boolean algebra and the Boolean difference, the author carefully and systematically develops and extends these concepts to subsume such diverse areas as two-level minimization, hazard detection, unate functions, fault diagnosis, functional decomposition, and many others. A significant part of this theory derives directly from previous work by the author and his colleagues at the Philips Research Laboratory.

The elegance of the underlying theory, together with its breadth of coverage and the clarity of the author's presentation, is destined to make this book a classic in its field.

Syracuse, New York, U.S.A. Sheldon B. Akers.

Contents.

1. Introduction

In 1954, in two papers dealing with coding theory and with error detection in data transmission, Reed [1954] and Muller [1954] use for the first time the concept of *Boolean derivative* or of *Boolean difference*. In 1958, Huffman uses the Boolean derivative and the connected concept of *Boolean Jacobian* to state a solvability criterion for a system of simultaneous Boolean equations. We may consider however that the *Boolean calculus of differences* (or *Boolean differential calculus*) is developed initially with the papers by Akers [1959] and by Talantsev [1959]. In his paper Akers shows how the concept of Boolean difference of the function f with respect to the variable x, difference that we denote $\Delta f/\Delta x$ and that we define as :

$$\frac{\Delta f}{\Delta x} = f(x) \oplus f(x \oplus 1) \qquad (1.1)$$

($\oplus$ represents the modulo-2 sum), allows us to study several properties of the Boolean function f. In particular Akers discovers that this concept leads to expansions of Boolean functions similar to the classical *Taylor-Maclaurin expansions*. He shows also that the concept of Boolean difference may be considered as an adequate mathematical tool in the problems of *function decomposition* and *solution of equations*. From the years 1967-1969 the Boolean difference received a considerable interest and numerous works were performed in the United States as well as in Europe in order to examine its domain of application. Amar and Condulmari [1967], Sellers, Hsiao and Bearnson [1968] and Davio and Piret [1969] use the Boolean difference to state algorithms able to solve a classical problem in switching theory, namely the detection of stuck-faults that may occur in logical networks. The work of these authors may be considered as a pioneering work in this domain : from 1970 the number of papers and of symposia dealing with the application of Boolean difference to the *diagnosis theory* increased considerably. The mean reason for this growth is due to the interest of the theory of diagnosis and to its immediate formulation in terms of Boolean differences. The author (Thayse [1970,1971]) introduces the notion of *Boolean differential* that we denote df and that we define as follows for a function $f(x)$:

$$df = f(x) \oplus f(x \oplus dx). \qquad (1.2)$$

This concept constitutes a generalization of that of derivative or of difference in the sense that the argument $(x \oplus 1)$ in (1.1) is replaced by $(x \oplus dx)$, dx being a function of the variable x : it will be called the *differential of* x. This generalization allows us to extend the domain of application of the difference to other classical problems in switching theory such as e.g. hazard detection (Thayse [1970, 1974e, 1975a] , Deschamps and Thayse [1973e]) the transformation between synchronous and asynchronous networks (Thayse [1972b]) and the detection of faults by means of variational tests (Thayse [1972a]). One may finally insert the concepts of *Boolean difference* and of *Boolean differential* in a theory called at one time *Boolean differential calculus* and at another time *Boolean calculus of differences* (Thayse and Davio [1973]). This theory constitutes a complete mode of representation of the algebra of Boolean functions of n variables (n =integer $\geqslant 1$) ; it allows a very simple interpretation of a large number of problems occurring a switching theory. In addition to the classical applications already quoted, let us mention also the minimization of Boolean polynomials (Bioul and Davio [1972], Thayse [1974a]), the Fourier or spectral analysis of Boolean functions (Lechner [1971], Davio, Thayse and Bioul [1972]) and a new approach to the decomposition of Boolean functions (Thayse [1972c]).

As it was established, mainly between the years 1970 and 1974, the Boolean differential calculus constitutes a self-contained theory for representing Boolean functions ; it presents several connections with the classical differential calculus and the calculus of finite differences. Let us also mention the fact that the Boolean differential calculus leads to the study of a differential calculus for the n-variable functions taking their values in the field of integers modulo a prime number p (Galois functions) (Thayse and Deschamps [1973] , Thayse [1974b], Davio, Deschamps and Thayse [1978], Pradhan [1978]). In a parallel way, a calculus of differences for functions taking their values in the ring of integers modulo an integer m was developed (Gazalé [1959], Tosic [1972], Deschamps and Thayse [1973], Kodandapani [1976], Benjauthrit and Reed [1976]). Davio, Deschamps and Thayse show that the differential calculus and the calculus of differences coincide for Boolean functions (Thayse and Deschamps [1973], Davio, Deschamps and Thayse [1978]). This coincidence between the two calculi justifies the fact that the function (1.1) has been called *Boolean derivative* by some authors and *Boolean difference* by others.

In 1961, Fadini studies Boolean operators and, among others, an operator denoted D_x and defined as follows :

$$D_x = f(x) \vee f(x \oplus 1) \quad . \qquad (1.3)$$

The author (Thayse [1973a]), starting from the work by Fadini, studies the functions

$$f \; \mathrm{T} \; f(x \oplus 1) \qquad (1.4)$$

and

$$f \; \mathrm{T} \; f(x \oplus dx) \qquad (1.5)$$

where T represents either the Boolean disjunction $\vee$ or the Boolean conjunction $\wedge$. In this paper we show that the function (1.4) is related to the classical concepts of *prime implicants* and of *prime implicates* of a Boolean function through the intermediate of simple formulas. In particular, we deduce from the function (1.4) a family of algorithms allowing us to obtain the set of prime implicants and the set of prime implicates of a Boolean function. Among other results we show (Thayse [1978]) that the consensus theory (Tison [1965, 1967, 1971]) may easily be formulated in terms of the function (1.4) and that the theory of fuzzy functions (Zadeh [1965],

Kauffman [1973]) may, at least partially, be formulated in terms of the function (1.5). In particular we construct (Davio and Thayse [1973]) from the function (1.5) an algorithm allowing us to detect the prime implicants of fuzzy functions.

In summary the functions (1.4, 1.5) generate a calculus on the Boolean functions very close under several aspects to the Boolean differential calculus ; it presents a definite interest in view of its applications quoted above. This calculus has been extended (Thayse [1976]) to functions taking their values in the ring of integers modulo an integer m.

The considerations relative to functions (1.1-1.5) suggest us to define a very general form of the Boolean difference, namely :

$$f(x) \uparrow f(\phi(x)) \ . \qquad (1.6)$$

In expression (1.6) "$\uparrow$" is a composition law to be defined further on and $\phi(x)$ is a function of x.

The main purpose of the present text is the study of difference functions issued from the form (1.6). For particular values of the law "$\uparrow$" and of the function ϕ , the function (1.6) represents any one of the functions (1.1-1.5). We shall show (see chapter 3) that other functions such as the *lower* and *upper envelopes* defined by Kuntzmann [1975], the *divided differences* or *test functions* used among others by Lapscher [1972] and the *sensitivity functions* defined by Sellers, Hsiao and Bearnson [1968] may also be represented in the form (1.6) for a given law $\uparrow$ and an appropriate function ϕ . This shows the importance taken by the study of the functional properties of the function (1.6). The work scheme adopted in the present text may be summarized as follows. We first study the most general functional properties of the function (1.6) by keeping the law $\uparrow$ and the function ϕ undetermined as long as possible. After that, more particular functional properties of the function (1.6) are studied by fastening firstly either the law $\uparrow$, or the function ϕ, and secondly by fastening simultaneously both the law $\uparrow$ and the function ϕ .

The results obtained by the functional study of (1.6) are principally of two types. First we show that a set of definitions, theorems and algorithms that are known for only one or for some of the functions (1.1-1.5) may be applied to a much larger set of functions. Essentially this represents generalization of known results. We then show how separate results obtained from the study of distinct problems reduce to unique statements. We obtain here a unified theory for known results.

The functional study of (1.6) and its resulting theorems are gathered in chapter 3. We present in this chapter together with the main known results of the Boolean calculus of differences a system of new theorems and properties in a form which is simultaneously condensed and generalized. It is to be noted that we introduce in this chapter only notions which may be interpreted in terms of switching networks. We present also in this chapter a first survey of applications of the concepts and of the theorems introduced in the successive sections ; these applications are extensively developed in chapter 4.

In a simplified way, we may characterize, the Boolean calculus of differences, as a set of closed formulas and of equalities. This type of relations also exist e.g. in the classical differential calculus. But the aim of the Boolean calculus of differences lies elsewhere ; it must be recalled that the logical design of switching circuits is the process of interconnecting logic modules to perform a specific task. Thus switching theory is concerned with establishing the optimal interconnection paths of logical elements required to meet a particular functional input/output specification. In this respect one of the main purposes of the Boolean calculus of differences is to solve problems ; but solving problems necessarily requests a *computation algorithm* : the latter may be considered as an intermediate between a *theorem* and the solution of a *problem*. Observe also that a theorem generates a system of several algorithms each of which may be used to solve a particular problem ; on the other hand the theorem may be considered as the mathematical support of a large class of problems. It is this aspect of the Boolean calculus of differences that we illustrate in chapter 4. This chapter gathers in a synoptic form the algorithms deriving from the theory developed in chapter 3 ; it describes the application of these algorithms to classical problems in switching theory.

Chapter 2 is an introductory chapter where the notations and the scope of the present work are presented.

Note finally that the theory is illustrated by means of an elementary example developed in sections 3.1.14 and 3.2.18 while a more elaborated example is dealed with in the course of chapter 4.

While writing this work I have contracted important debts of gratitude. I wish to thank first Professor Vitold Belevitch, Director of the Philips Research Laboratory, both for his constant interest and encouragement and for giving me the possibility of materializing this work within the framework of the Philips Research Laboratory activities. A part of this work was presented as a Ph. D. Thesis at the chaire des systèmes logiques of the Ecole Polytechnique Fédérale de Lausanne (Switzerland).

In this respect special thanks are due to the Professors Daniel Mange and Jacques Zahnd. I am also indebted to many of my colleagues of the Research Laboratory but particularly to Mr Claude Fosséprez Director at Philips telecommunication Industry and to Dr. Marc Davio group leader who initiated me to Boolean algebra and switching theory. Moreover Dr. Marc Davio suggested many improvements for the present work. Dr. Philippe Delsarte (Philips Research Laboratory) carefully read the manuscript and suggested numerous corrections. Professor Sheldon B. Akers (University of Syracuse, New York, and General Electric) kindly accepted to preface this book. Mrs Moës typed the manuscript with amability and competence.

Brussel, October 1980 André Thayse

2. Canonical expansions of Boolean functions

2.1. General Boolean notations

2.1.1. Definitions and elementary properties

A *Boolean algebra* is an algebraic structure

$$< B, \vee, \wedge, \bar{\ } >$$

where B is a set (the *carrier* of the algebra) and in which two binary operations called *disjunction* ($\vee$) and *conjunction* ($\wedge$) and one unary operation called *negation* or *complementation* ($\bar{\ }$) are defined. Roughly speaking, these laws have the same properties as the set-theoretical union, intersection and complementation of subsets of a fixed space ; they are characterized by a set of axioms. Many equivalent sets of axioms characterizing $\vee, \wedge, \bar{\ }$ are known; we assume here the following one :

$\forall a,b,c \in B$: ,

(A_1) $a \vee b = b \vee a$, $a \wedge b = b \wedge a$;

(A_2) $a \vee (b \vee c) = (a \vee b) \vee c$, $a \wedge (b \wedge c) = (a \wedge b) \wedge c$;

(A_3) $a \vee (a \wedge b) = a$, $a \wedge (a \vee b) = a$;

(A_4) $a \wedge (b \vee c) = (a \wedge b) \vee (a \wedge c)$, $a \vee (b \wedge c) = (a \vee b) \wedge (a \vee c)$;

(A_5) $(a \wedge \bar{a}) \vee b = b$, $(a \vee \bar{a}) \wedge b = b$.

It is the system of axioms proposed by Sikorski [1969] p.3.
From these axioms we deduce several properties such as e.g. the two following ones :

$$\forall a,b \in B : \quad a \vee a = a \, , \quad a \wedge a = a \; ; \tag{2.1}$$

$$a \vee b = b \Leftrightarrow a \wedge b = a \, . \tag{2.2}$$

The essential property that we deduce from these axioms is that they allow us to define two dual partial ordering relations on B. Let us recall(see e.g. Grätzer [1971] ,p.1) that partial ordering relation on a set B is a relation which is at the same time reflexive, antisymmetric and transitive ;

$\forall a,b,c \in B$ we have :

(P_1) a R a (reflexivity) ;

(P_2) a R b and b R a imply a=b (antisymmetry) ;

(P_3) a R b and b R c imply a R c (transitivity) ;

<u>Proposition</u> *The dual relations*

$$R = \{(a,b) \mid a \vee b = b\} \quad \text{and} \quad R_d = \{(a,b) \mid a \vee b = a\}$$

are partial ordering relations

Let us denote R as $\leqslant$ and R_d as $\geqslant$ and let us adopt the corresponding terminology (Grätzer [1971], p.2,3). Then :

$$\forall c,d \in B : c \vee d = \textit{supremum}\ (c,d)\ ;\ c \wedge d = \textit{infimum}\ (c,d).$$

From this proposition we deduce the existence of *minimum* and *maximum* elements : the proposition together with the axiom (A_5) show that :

$$\forall a,b,c \in B : a \wedge \bar{a} \leqslant b\ ;\ c \wedge \bar{c} \leqslant d.$$

Choosing $b=c \wedge \bar{c}$ and $d=a \wedge \bar{a}$, we obtain by antisymmetry :

$$\forall a,c \in B : a \wedge \bar{a} = c \wedge \bar{c}\ .$$

The minimum element discovered will be written $\underline{\underline{0}}$. The corresponding maximum element (equal to $a \vee \bar{a}$) will be written $\underline{\underline{1}}$.

The proposition contains the statement of the *duality principle* (Grätzer [1971], p.7; Rudeanu [1974], p.5). This principle may be stated as follows :

Let a property of an arbitrary Boolean algebra be expressed in terms of the operations $\vee, \wedge, \bar{\ }$ *and the elements* $\underline{0}$ *and* $\underline{1}$. *Then the "dual property" obtained by interchanging* $\vee$ *with* $\wedge$ *and* 0 *with* 1, *also holds.*

In the following section we introduce a general Boolean notation ; it has the property of gathering in a single expression any pair of dual relations.

2.1.2. Boolean algebra

Denote by $\top$ either the operation of disjunction or of conjunction; the second of these operations will then be denoted by $\perp$. The writing of the axioms (A_1)-(A_5), of the properties (2.1) and (2.2) and of the proposition 2.1.1 becomes :

$\forall \perp, \top \in \{\vee, \wedge\},\ \top \neq \perp\ ;\ \forall a,b, \in B$:

(A_1) $a \top b = b \top a$

(A_2) $a \top (b \top c) = (a \top b) \top c$;

(A_3) $a \top (a \perp b) = a$;

(A_4) $a \top (b \perp c) = (a \top b) \perp (a \top c)$;

(A_5) $(a \top \bar{a}) \perp b = b$.

Moreover :

$$\forall a,b \in B : \qquad a \top a = a\ ; \qquad (2.3)$$

$$a \top b = b \Leftrightarrow a \perp b = a\ . \qquad (2.4)$$

Proposition. *The relation*

$$R = \{(a,b) \mid a \top b = b\}$$

is a partial ordering relation; the elements $a \top b$ *and* $a \perp b$ *are the supremum and the infimum of the pair* $\{a,b\}$ *respectively.*

If we want to develop a theory using these general notations, we have to solve the difficulty coming from the representation of the extreme : we may no longer use the elements $\underline{\underline{0}}$ and $\underline{\underline{1}}$ as such. The particular form of the axiom (A_5) of this section suggests us to choose $a \top \bar{a}$ and this choice is appropriate since :

(a) $a \top \bar{a}$ is *unit* with respect to $\top$:

$\forall b \in B : (a \top \bar{a}) \top b = a \top \bar{a}$;

(b) $a \top \bar{a}$ is *zero* with respect to $\perp$:

$\forall b \in B : (a \top \bar{a}) \perp b = b$.

Adopting a classical notation for the zero element (also classically referred to as the *neutral element*) we shall represent $a \top \bar{a}$ by $e_\perp$. The two ways of considering the Boolean algebra are then condensed in the following form

$$< B, \top, e_\top, \perp, e_\perp, \bar{\ } > \tag{2.5}$$

where each of the operations $\top$ and $\perp$ is followed by its zero element. It becomes then elementary to formulate, in the general frame defined by (2.5), the usual properties of Boolean algebras. Let us illustrate this by proving the property :

$$(a,b) \in R \Leftrightarrow \bar{a} \top b = e_\perp \ . \tag{2.6}$$

Assume first that $(a,b) \in R$. By definition we have $a \top b = b$, i.e. : $\bar{a} \top (a \top b) = \bar{a} \top b$. Owing to the associativity, the property $a \top \bar{a} = e_\perp$ and the unit character with respect to $\top$ of this element, the left side reduces to $e_\perp$. Assume then that $\bar{a} \top b = e_\perp$; thus $a \perp (\bar{a} \top b) = a \perp e_\perp$ and this gives, by distributivity and by taking the zero character of $e_\perp$: $(a \perp \bar{a}) \top (a \perp b) = a$, $e_\top \top (a \perp b) = a$, and finally by (2.2) $a \top b = b$, $(a,b) \in R$.

In that context the Boolean absorption law and the De Morgan law are written respectively :

$$a \top (\bar{a} \perp b) = a \top b \ , \tag{2.7}$$

$$\overline{a \top b} = \bar{a} \perp \bar{b} \ . \tag{2.8}$$

2.1.3. Boolean ring

A *Boolean ring* is an algebraic structure

$$< B, \oplus, 0, ., 1 >$$

with an additive law ($\oplus$) with zero element 0, a multiplicative law (.) with zero element 1 and the usual axioms of the rings :

$\forall a,b,c \in B$:

(B_1) $a \oplus b = b \oplus a$;

(B_2) $a \oplus (b \oplus c) = (a \oplus b) \oplus c$;

(B_3) The equation $a \oplus x = b$ has a unique solution for x ;

(B_4) $a.(b.c) = (a.b).c$;

(B_5) $a.(b \oplus c) = (a.b) \oplus (a.c)$;

(B_6) $(a \oplus b).c = (a.c) \oplus (b.c)$;

We have moreover :

(B_7) $a.a = a \quad \forall\, a \in B.$

A Boolean ring satisfies the following properties :

$$\forall a,b \in B : a \oplus a = 0 \; ; \tag{2.9}$$

$$a \oplus b = 0 \Rightarrow a = b \; ; \tag{2.10}$$

$$a \, . \, b = b \, . \, a \quad . \tag{2.11}$$

The central property is the one to one correspondence between a Boolean algebra and a Boolean ring. We first state this property, due to Stone [1935] for the Boolean algebra (2.5). Then we come to the two dual ways of presenting a Boolean algebra, as they were described by Rudeanu [1961] .

<u>Theorem.</u> *The Boolean algebra* (2.5) *is a Boolean ring*

$$< B, \oplus, e_T, \perp, e_\perp > \tag{2.12}$$

if we define the additive law $\oplus$ *as follows :*

$$a \oplus b = (a \perp \bar{b}) \; T \; (\bar{a} \perp b) \tag{2.13}$$

Conversely, the Boolean ring (2.12) *is a Boolean algebra if we define the operation* T *and the complementation* $\bar{}$ *by :*

$$a \; T \; b = a \oplus b \oplus (a \perp b) \; , \tag{2.14}$$

$$\bar{a} = a \oplus e_\perp \quad . \tag{2.15}$$

<u>Proof.</u>

We verify first that if (2.5) is a Boolean algebra, then (2.12) and (2.13) define a Boolean ring. The properties (B_4) and (B_7) are equivalent to the properties (A_2) and (2.3) of section 2.1.2 respectively; (B_1), (B_2), (B_5) and (B_6) may be shown by means of elementary computations. For example the proof of (B_5), i.e. :

$$a \perp (b \oplus c) = (a \perp b) \oplus (a \perp c)$$

reduces by (2.13) to :

$$a \perp ((b \perp \bar{c}) T (\bar{b} \perp c)) = ((a \perp b) \perp \overline{(a \perp c)} \, T (\overline{(a \perp b)} \perp (a \perp c)).$$

Elementary algebraic manipulations allow us to reduce the two members of this equality to the form $(a \perp b \perp \bar{c}) T (a \perp \bar{b} \perp c)$. To prove ($B_3$), it is sufficient for example to show that e_T is neutral for $\oplus$ i.e. :

$$a \oplus e_T = a \quad , \tag{2.16}$$

and that each element is its own symmetric. The unicity of the symmetric elements and of the solution $a \oplus x = b$ follow from these properties. We may quote a similar type of proof for stating the converse of this property. □

We easily verify that if $\perp$ is the conjunction $\oplus$ is nothing but the *modulo-2 sum*

(denoted ⊕) while if ⊥ is the disjunction, ⊕ is the *identity* (denoted ⊙).
It is useful to state expressions for the neutral elements in terms of the ring operations. We verify that :

$$e_{⊕} = e_T = a \perp \bar{a} = a\perp (a \oplus e_\perp) = (a \perp a) \oplus (a \perp e_\perp) = a \oplus a \quad , \tag{2.17}$$

$$e_{⊙} = e_\perp = a \top \bar{a} = a \odot \bar{a} \quad . \tag{2.18}$$

2.1.4. The two-element Boolean algebra ; Boolean functions.

The simplest Boolean algebra is the *two-element Boolean algebra* ; we take for B the integers 0 and 1 and we denote by B_2 the set {0,1}. Observe also that in the dual form (2.5) the set {0,1} is described as {0 ⊥ 1, 0 T 1}.

The importance of the two-element Boolean algebra derives both from its particular role in the general theory of Boolean algebras and to the number of its applications in practical problems ; the Boolean algebra B_2 will be the only algebra considered in the present text.

We call *Boolean mapping* or, according to a usual terminology, *Boolean function*, any mapping :

$$f : B_2^n = B_2 \times B_2 \times \ldots \times B_2 \to B_2 \quad .$$

i.e. a mapping with arguments and values in B_2 (here "×" means the Cartesian product). The value of f at the value $(x_{n-1},\ldots,x_1,x_0) \in \{0,1\}^n$ is denoted $f(x_{n-1},\ldots,x_1,x_0)$.

2.1.5. Well-formed Boolean expression ; well-formed Galoisian expression.

The Boolean functions $f : B_2^n \to B_2$ may be represented either by means of *tables* (truth tables) or by means of *algebraic expressions*. Further on we shall consider two kinds of algebraic expressions, namely *well-formed Boolean expressions* and *well-formed Galoisian expressions* (see Rudeanu [1974] ,p.17, Hammer and Rudeanu [1968] , p.8, Preparata and Yeh [1973] , p. 239).

Let $\underline{x}=(x_{n-1},\ldots,x_1,x_0)$ be a set of symbols called *variables*. *Well-formed Boolean expressions* are defined inductively as follows :

1. $0,1,x_i$ $(i=0,1,\ldots,n-1)$ are well-formed Boolean expressions ;
2. If E_0 and E_1 are well-formed Boolean expressions, then $(E_0 \top E_1)$, $(E_0 \perp E_1)$ and $(\bar{E}_0)$ are well-formed Boolean expressions ;
3. No other well-formed Boolean expressions exist that cannot be generated by a finite number of applications of rules (1) and (2).

We are able to write a Boolean expression comprising, beside literals and T,⊥ ,¯, nested symmetric pairs of parentheses, ( ), enclosing the expressions operated upon according to rule (2) of the preceding definition. For example, if E_0 and E_1 are expressions, then $E_0 \top E_1$ is written as $(E_0 \top E_1)$. Therefore for some literals x_0,x_1,x_2,x_3

$$((x_0 \top x_1 \top ((x_2 \top x_3) \perp ((\overline{x_1 \perp x_2})))) \perp (x_2 \top (x_2 \perp x_3)))$$

is an expression written according to the given rule. Since it may be difficult, especially in complicated cases, to spot which parenthesis symbols form a symmetric pair, simplification rules are common in the use of parentheses if no ambiguity results. For example, in any well-formed Boolean expression *complementation* is performed before *conjunction* and *conjunction* before *disjunction* in accordance with the so-called *precedence rule*. Remember that in ordinary algebraic expressions we have the precedence rule : multiplication-addition. Note also that the above precedence rules are used in the Boolean expressions where the operations T and ⊥ are replaced either by the disjunction, or by the conjunction and by its dual operation. Moreover, as in ordinary algebra it is of common use to drop the multiplication symbol ".", in the same way we shall generally omit the conjunction symbol "∧" in the Boolean expressions.

Boolean functions may also be represented by means of (well-formed) *Galoisian expressions* defined as follows :

1. $0, 1, x_i$ or $\bar{x}_i$ $(i=0,1,\dots,n-1)$ are Galoisian expressions ;
2. If E_0 and E_1 are Galoisian expressions, then $(E_0 \oplus E_1)$ and $(E_0 \top E_1)$ are Galoisian expressions.
3. No other Galoisian expression exist that cannot be generated by a finite number of applications of rules 1 and 2.

We are now able to write a Galoisian expression as a string of symbols comprising literals $0, 1, x_i$ or $\bar{x}_i$, operation symbols $\oplus$, T (the latter one being generally omitted), and parentheses. The parentheses may again partially be dropped by imposing the precedence rule T- $\oplus$. It is easy to see that Boolean expressions and Galoisian expressions are particular computation schemata allowing us to evaluate the local values $f(x_{n-1},\dots,x_1,x_0)$: it is assumed that the values on x_i are evaluated according to the laws defined in $B_2=\{0,1\}$. We shall say, according to a well-established use, that the n-variable Boolean function $f(\underline{x})$ has that expression as *expansion* . The main purpose of sections 2.2 and 2.3 is to study the *canonical expansions* of Boolean functions. By the calling *canonical* we mean at the same time the universal character of the computation scheme, i.e. its applicability to any function, and its unicity. In section 2.2 we will restrict ourselves to expansions with respect to one variable ; the expansions with respect to several variables will be presented in section 2.3.

2.2. Partial expansions with respect to one variable.

2.2.1. Lagrange expansion and redundant expansion.

Let x be an element of $\underline{x}$ (when dealing with only one variable the index "i" in x_i will be omitted).

Let us first consider the functions $f(x)$ of the unique variable x.

<u>Theorem</u> (*Lagrange expansion*)

Any Boolean function $f(x)$ has the expansion :

$$f(x) = (\bar{x} \top f(e_\perp)) \perp (x \top f(e_\top)) \ . \tag{2.19}$$

<u>Proof</u>

The left side and right side parts of (2.19) are identical for both $x=e_\perp$ and $x=e_\top$. For example, for $x=e_\perp$ the left side part is equal to $f(e_\perp)$, while the right side part may be written, in view of $\bar{e}_\perp = e_\top$:

$$\begin{aligned}(e_\top \top f(e_\perp)) \perp (e_\perp \top f(e_\top)) &= f(e_\perp) \perp e_\perp \ ,\\ &= f(e_\perp) \ .\end{aligned}$$

□

<u>Corollary</u>

The Lagrange expansion (2.19) may be written in the following equivalent forms :

$$\begin{aligned} f(x) &= ((x \textcircled{\top} e_\top) \top f(e_\perp)) \perp ((x \textcircled{\top} e_\perp) \top f(e_\top)) \\ &= ((x \textcircled{\perp} e_\perp) \top f(e_\perp)) \perp ((x \textcircled{\perp} e_\top) \top f(e_\top)) \end{aligned} \tag{2.20}$$

$$\begin{aligned} f(x) &= ((x \textcircled{\top} 1) \top f(0)) \perp ((x \textcircled{\top} 0) \top f(1)) \\ &= ((x \textcircled{\perp} 0) \top f(0)) \perp ((x \textcircled{\perp} 1) \top f(1)) \end{aligned} \tag{2.21}$$

<u>Proof</u>

Owing to (2.15) and (2.16), the expansion (2.19) may be written in the forms (2.20) respectively; the expansions (2.21) are obtained in the same way. □

The equivalent forms (2.19-2.21) of the Lagrange expansion are dual expressions for the *canonical conjunctive* and *disjunctive expansions*. For the n-variable functions $f(\underline{x})$, the expansions (2.19-2.21) are *partial expansions* with respect to a single variable x.

Starting from (2.19) and using the distributivity of $\perp$ with respect to $\top$, we obtain the following expansion after interchanging $\top$ and $\perp$:

$$f(x) = (\bar{x} \top f(e_\perp)) \perp (x \top f(e_\top)) \perp (f(e_\perp) \top f(e_\top)) \tag{2.22}$$

We shall call (2.22) the *redundant canonical expansion* since the term $(f(e_\perp) \top f(e_\top))$ is redundant. Forms equivalent to (2.22) could also be obtained by starting from the equivalent expansions (2.20) or (2.21).

The expansions (2.19) and (2.22) are called : *latticial canonical expansions* ; they are based on the use of the laws $\top$ and $\perp$ which, in view of proposition 2.1.2, are latticial composition laws. In section 2.2.3 we shall consider *Galoisian expansions* : they use the laws $\textcircled{\top}$ and $\top$ of the Boolean ring and either x or $\bar{x}$ (or an equivalent form of these variables). The expansions of section 2.2.2. are hybrid forms : they use the ring operations $\textcircled{\top}$ and $\top$ and contain simultaneously the forms x and $\bar{x}$

of the variable.

2.2.2. Hybrid expansions

Theorem

Any Boolean function f(x) *has the expansions :*

$$f(x) = (\bar{x}\ T\ f(e_\perp))\ \textcircled{T}\ (x\ T\ f(e_\top))\ , \tag{2.23}$$

$$f(x) = (\bar{x}\ T f(e_\perp)\ T\ \bar{f}(e_\top))\textcircled{T}(x\ T\ f(e_\top) T\bar{f}(e_\perp))\textcircled{T}(f(e_\perp) T f(e_\perp))\ . \tag{2.24}$$

Proof

Taking into account (2.19) and (2.14) we may write :

$$f(x) = (\bar{x}\ T\ f(e_\perp))\ \textcircled{T}\ (x\ T\ f(e_\top))\ \textcircled{T}\ (\bar{x}\ T\ f(e_\perp)\ T\ x\ T\ f(e_\top))\ .$$

We deduce the expansion (2.23) from the preceding relation by taking into account that $\bar{x}\ T\ x = e_\perp$ is unit for T and zero for $\textcircled{T}$. The expansion (2.24) is deduced from (2.22) as (2.23) was deduced from (2.19). □

2.2.3. Galoisian expansions

Theorem

Any Boolean function $f(\underline{x})$ has the expansions

$$f(x) = f(e_\perp)\ \textcircled{T}\ (x\ T\ (f(e_\perp)\ \textcircled{T}\ f(e_\top)))\ , \tag{2.25}$$

$$f(x) = f(e_\top)\ \textcircled{T}\ (\bar{x}\ T\ (f(e_\perp)\ \textcircled{T}\ f(e_\top)))\ . \tag{2.26}$$

Proof

Starting from relation (2.23) and replacing $\bar{x}$ by $x\ \textcircled{T}\ e_\top$, we successively obtain :

$$\begin{aligned} f(x) &= ((x\ \textcircled{T}\ e_\top)\ T\ f(e_\perp))\ \textcircled{T}\ (x\ T\ f(e_\top))\ , \\ &= (x\ T\ f(e_\perp))\ \textcircled{T}\ (e_\top\ T\ f(e_\top))\ \textcircled{T}\ (x\ T\ f(e_\top)),\ \text{(by distributivity)}\ , \\ &= f(e_\top)\ \textcircled{T}\ (x\ T\ (f(e_\perp)\ \textcircled{T}\ f(e_\top)))\ .\ \ \text{(by grouping terms).} \end{aligned}$$

Relation (2.26) may be proved in a similar way. □

In view of the equalities (2.15) and (2.16), i.e. :

$$x = x\ \textcircled{T}\ e_\perp \quad \text{and}\ \bar{x} = x\ \textcircled{T}\ e_\top\ ,$$

the polynomial expansions (2.25) and (2.26) may be gathered into the following unique expression :

$$f(x) = f(h)\ \textcircled{T}\ ((x\ \textcircled{T}\ h)\ T\ (f(h)\ \textcircled{T}\ f(\bar{h})))\ ,\ h \in \{e_\top, e_\perp\}\ . \tag{2.27}$$

The polynomial expansions (2.27) are known as the *Newton expansions* or also as the *Taylor expansions ;* for $\textcircled{T}$ the modulo-2 sum, expansions (2.27) are also called *Reed-Muller expansions*.

2.2.4. Matrix expression of the canonical expansions.

It will often be useful to put the preceding expansions in a matrix form ; let us designate by [AM] the matricial product using the additive law A and the multiplicative law M. This notation allows us to write the expansions (2.19, 2.22, 2.25, 2.26) in the following matrix form.

$$[f(e_\perp),\ f(e_\top)][\perp\top]\begin{bmatrix}\bar{x}\\ x\end{bmatrix} \qquad (2.28)$$

$$[f(e_\perp),f(e_\top),f(e_\perp)\top f(e_\top)][\perp\top]\begin{bmatrix}\bar{x}\\ x\\ e_\top\end{bmatrix} \qquad (2.29)$$

$$[f(e_\perp),f(e_\top),f(e_\perp)\textcircled{\top}f(e_\top)]\ [\textcircled{\top}\top]\begin{bmatrix}e_\top & e_\perp\\ e_\perp & e_\top\\ x & \bar{x}\end{bmatrix} \qquad (2.30)$$

The matrix product (2.30) gathers the expansions (2.25) and (2.26).

2.2.5. Equivalent forms for canonical expansions

We expressed the expansions described in section 2.2 in terms of the neutral elements $e_\top$ and $e_\perp$. It is often useful to obtain the canonical expansions in terms of 0 and 1.

Consider a function f(x) described by means of a well-formed expression g(x) satisfying the identity :

$$g(x,\ e_\top,e_\perp) = g(x,e_\perp,e_\top)\ . \qquad (2.31)$$

This expression (2.31) means that g does not change when $e_\top$ and $e_\perp$ are interchanged. Since $\bar{e}_\top = e_\perp$ any expression g(x) representing f(x) and satisfying (2.31) satisfies also the relation (2.32) :

$$f(x) = g(x,0,1) = g(x,1,0)\ . \qquad (2.32)$$

In order to obtain the canonical expansions of a function in terms of the elements 0 and 1, it is sufficient to obtain canonical expansions symmetric in $e_\top$ and $e_\perp$ and then to apply the substitution suggested by (2.32). For example, the application of this rule allows us to deduce the expansions (2.21) from (2.20). Let us give some other examples. Taking into account (2.15) and (2.16), the expansion (2.22) may be written :

$$f(x) = ((x \textcircled{\top} e_\top)\top f(e_\perp)) \perp ((x \textcircled{\top} e_\perp)\top f(e_\top)) \perp (f(e_\perp)\ \top\ f(e_\top))\ , \qquad (2.33)$$

which is a symmetric form in $e_\top$ and $e_\perp$; we deduce the expansion :

$$f(x) = ((x \circledT 1) \top f(0)) \perp ((x \circledT 0) \top f(1)) \perp (f(0) \top f(1)) . \quad (2.34)$$

From expansions (2.25) and (2.26) we deduce :

$$\begin{aligned} f(x) &= f(e_\perp) \circledT ((x \circledT e_\perp) \top (f(e_\perp) \circledT f(e_\top))) \\ &= f(e_\top) \circledT ((x \circledT e_\top) \top (f(e_\top) \circledT f(e_\perp))) \quad . \end{aligned}$$

Hence the Galoisian expansions may also be written in the following forms :

$$\begin{aligned} f(x) &= f(0) \circledT ((x \circledT 0) \top (f(0) \circledT f(1))) , \\ &= f(1) \circledT ((x \circledT 1) \top (f(0) \circledT f(1))) , \quad (2.35) \\ &= f(h) \circledT ((x \circledT h) \top (f(0) \circledT f(1))) , \quad h \in \{0,1\} \end{aligned}$$

2.3. Complete expansions

2.3.1. Lagrange expansions.

The Boolean function expansions that we consider in this section are obtained by iteration on the set of variables $x_i \in \underline{x}$ of the rules considered in section 2.2. Let us first consider the Lagrange expansions. Be E a set formed by two complementary zero elements, i.e. : $E = \{0,1\}$ or $E = \{e_\top, e_\perp\}$. The forms (2.20, 2.21) of the Lagrange expansion may be written :

$$f(x) = \perp_e ((x \circledperp e) \top f(e)) , \; e \in E . \quad (2.36)$$

If in (2.36) we substitute x_0 for x and if we expand the functions $f(x_0=e)$, $e \in E$, successively according to the n-1 remaining variables of $\underline{x}$, we obtain the *Lagrange canonical expansion* of an n-variable Boolean function , i.e. :

$$f = \perp_{\underline{e}} [(\top_{i=0,n-1} (x_i \circledperp e_i)) \top f(\underline{e})] , \; \underline{e} = (e_{n-1}, \ldots, e_1, e_0), e_i \in E \;\; \forall i \quad . \quad (2.37)$$

For $E = \{0,1\}$ we obtain the matrix form of (2.37) by using the concepts of *state vector* and of *Kronecker matrix product*. The state vector, that we denote by $[f_{\underline{e}}]$, of an n-variable function $f(\underline{x})$ is a (1×2^n) matrix : $[f(e_{n-1}, \ldots, e_1, e_0)]$; the element $f(e_{n-1}, \ldots, e_1, e_0)$ is at the $(2^0 e_0 + 2^1 e_1 + \ldots + 2^{n-1} e_{n-1})$-th place. The Kronecker matrix product (see e.g. Marcus and Minc [1964] , p.8 and Bellman [1960]); using the multiplicative law T will be denoted $\overset{\top}{\otimes}$; it is defined as follows.

If A is e.g. an $(m \times n)$ matrix of the elements a_{ij} and if B is a matrix of any dimension , then :

$$A \overset{\top}{\otimes} B = \begin{bmatrix} a_{11} \top B & a_{12} \top B & \ldots & a_{1n} \top B \\ a_{21} \top B & a_{22} \top B & \ldots & a_{2n} \top B \\ \vdots & \vdots & & \vdots \\ a_{m1} \top B & a_{m2} \top B & \ldots & a_{mn} \top B \end{bmatrix}$$

The concept of state vector, the Kronecker matrix product, and relation (2.37) allow us to write the formula (2.38) which is the extension of (2.21) to the n-variable canonical expansion.

$$f(\underline{x}) = [\,f_{\underline{e}}\,]\;[\perp \mathrm{T}\,]\;\Big(\bigotimes_{i=n-1,0}^{\mathrm{T}} \begin{bmatrix} x_i \odot 0 \\ \\ x_i \odot 1 \end{bmatrix}\Big)\;. \tag{2.38}$$

The matrix expression (2.38) is, at least formally, less general than the expansion (2.37) : this last expansion reduces to (2.38) for E= {0,1} . The main reason for restricting ourselves to the particular case E= {0,1} when using a matrix notation is that the lexicographic order : $2^0 e_0 + 2^1 e_1 + \ldots + 2^{n-1} e_{n-1}$, and hence the state vector are defined only for $e_i \in \{0,1\}$. This does not restrict the generality of the writing (2.38) : this relation being symmetric in 0,1 for any variable $x_i \in \underline{x}$ (in the sense of relation (2.32)), we may replace the 0 and the 1 by $e_\perp$ and e_T respectively for each of the variables x_i of $\underline{x}$. For a two-variable function $f(x_1,x_0)$, the relation (2.38) becomes :

$$f(x_1,x_0) = [\,f(0,0), f(0,1), f(1,0), f(1,1)\,][\perp \mathrm{T}]\;\Big(\begin{bmatrix} x_1 \odot 0 \\ \\ x_1 \odot 1 \end{bmatrix} \overset{\mathrm{T}}{\otimes} \begin{bmatrix} x_0 \odot 0 \\ \\ x_0 \odot 1 \end{bmatrix}\Big)\;;$$

this relation may also be written :

$$f(x_1,x_0) = [\,f(e_\perp,e_\perp), f(e_\perp,e_\mathrm{T}), f(e_\mathrm{T},e_\perp), f(e_\mathrm{T},e_\mathrm{T})\,][\perp \mathrm{T}]\;\Big(\begin{bmatrix} x_1 \odot e_\perp \\ \\ x_1 \odot e_\mathrm{T} \end{bmatrix} \overset{\mathrm{T}}{\otimes} \begin{bmatrix} x_0 \odot e_\perp \\ \\ x_0 \odot e_\mathrm{T} \end{bmatrix}\Big).$$

Further on for the sake of simplicity we shall restrict ourselves to expressions containing the zero elements 0 and 1 ; this does not reduce the generality of the proposed formulas.

2.3.2. The extended state vector

The extension to the whole set of variables of the redundant canonical expansions and of the polynomial expansions is easiest obtained by starting from the matrix formulations (2.29, 2.30). First we introduce the concept of *extended state vector*. This concept has been introduced in an elementary form by Davio [1971]; the author (Thayse [1976]) defines the extended state vector in the form developed below.

Tha partial *extended vector* with respect to a variable x_1 and according to the law $\uparrow$ ($\uparrow$ represents any binary law, i.e. : $\vee, \wedge, \oplus$ or $\odot$) is defined as follows:

$$\underline{\phi}_{x_1}^{(\uparrow)}(f) = [\,f_{x_1=0},\; f_{x_1=1},\; f_{x_1=1} \uparrow f_{x_1=0}\,]\,. \tag{2.39}$$

For a two-variable function, the *extended state vector* $\underline{\phi}^{(\dagger)}(f)$ is defined from (2.39) and from the recurrent expression (2.40) :

$$\underline{\phi}^{(\dagger)}(f(x_1,x_0)) = [\underline{\phi}_{x_0}^{(\dagger)}(f_{x_1=0}),\ \underline{\phi}_{x_0}^{(\dagger)}(f_{x_1=1}),\ \underline{\phi}_{x_0}^{(\dagger)}(f_{x_1=1} \dagger f_{x_1=0})] \qquad (2.40)$$

The extended state vector $\underline{\phi}^{(\dagger)}$ of an (n+1)-variable function is defined in a similar way by using the recurrent formula :

$$\underline{\phi}^{(\dagger)}(f(\underline{x},y)) = [\underline{\phi}_{\underline{x}}^{(\dagger)}(f(\underline{x},0)),\ \underline{\phi}_{\underline{x}}^{(\dagger)}(f(\underline{x},1)),\ \underline{\phi}_{\underline{x}}^{(\dagger)}(f(\underline{x},0) \dagger f(\underline{x},1))] \qquad (2.41)$$

The extended state vector of an n-variable function is a (1×3^n) matrix ; observe that the ordering in which the variables must be considered in (2.40, 2.41) is important, i.e. :

$$\underline{\phi}_{x_0}^{(\dagger)}(\underline{\phi}_{x_1}^{(\dagger)}(f)) \neq \underline{\phi}_{x_1}^{(\dagger)}(\underline{\phi}_{x_0}^{(\dagger)}(f)) .$$

The extended state vector is obtained by matrix multiplication from the state vector.

Let $\dagger$ be one of the operations $\vee,\wedge,\oplus$ or $\odot$ and let $\dagger_d$ be the operation which distributes $\dagger$; remember that $\vee$ and $\wedge$ are distributive for each other while $\wedge$ distributes $\oplus$ and $\vee$ distributes $\odot$. Be $e_\dagger$ and $e_{\dagger d}$ the respective zero elements for the laws $\dagger$ and $\dagger_d$; we have (see sections 2.1.2 and 2.1.3) :

$$\bar{e}_\dagger = e_{\dagger d} .$$

Moreover $e_\dagger$ is unit for the law $\dagger_d$; this allows us to write :

$$[f(0),f(1),f(0) \dagger f(1)] = [f(0),f(1)][\dagger\ \dagger_d]\begin{bmatrix} e_\dagger & e_{\dagger d} & e_\dagger \\ e_{\dagger d} & e_\dagger & e_\dagger \end{bmatrix} \qquad (2.42)$$

If M is the (2×3) matrix of the right side of (2.42) and if $\overset{\dagger}{\otimes} M^n$ means the Kronecker power of matrices :

$$\underbrace{M \overset{\dagger}{\otimes} M \overset{\dagger}{\otimes} \dots \overset{\dagger}{\otimes} M}_{n \text{ times}}$$

the following relation is an immediate consequence of the above definitions :

$$\underline{\phi}^{\dagger}(f) = [f_{\underline{e}}]\ [\dagger\ \dagger_d](\overset{\dagger}{\otimes}{}^d M^n) . \qquad (2.43)$$

2.3.3. Redundant canonical expansions

First of all we shall obtain the redundant canonical expansions of an n-variable function from the extended state vectors $\underline{\phi}^{(\vee)}(f)$ and $\underline{\phi}^{(\wedge)}(f)$.

Theorem

$$f = \underline{\phi}^{(T)}(f)[\perp T] \quad \left(\overset{T}{\underset{i=n-1,0}{\otimes}} \begin{bmatrix} x_i \textcircled{T}\, 0 \\ x_i \textcircled{T}\, 1 \\ e_{\textcircled{T}} \end{bmatrix} \right) . \tag{2.44}$$

Proof

The proof of formula (2.44) is obtained in an inductive way. For one-variable functions, the formulas (2.44) and (2.29) coincide after replacement in (2.29) first of $\bar{x}$ by $x \textcircled{T}\, e_\perp$, of x by $x \textcircled{T}\, e_\perp$ and of e_T by $e_{\textcircled{T}}$. In the obtained expression we replace then $e_\perp$ by 0 and e_T by 1. Assume that relation (2.44) holds true for the n-variable functions of $\underline{x} = (x_{n-1},\ldots,x_1,x_0)$; we may then expand the (n+1)-variable functions according to (2.44) and restricting ourselves to the first n variables, i.e. :

$$f(x_{n-1},\ldots,x_1,x_0,y) = \underline{\phi}^{(T)}_{x_0x_1\ldots x_{n-1}}(f)[\perp T] \quad \left(\overset{T}{\underset{i=n-1,0}{\otimes}} \begin{bmatrix} x_i \textcircled{T}\, 0 \\ x_i \textcircled{T}\, 1 \\ e_{\textcircled{T}} \end{bmatrix} \right) .$$

Each of the elements of $\underline{\phi}^{(T)}_{x_0x_1\ldots x_{n-1}}(f)$ being a function in one variable y, we may use the expansion (2.29) which allows us to write :

$$f(\underline{x},y) = \underline{\phi}^{(T)}(f)[\perp T] \quad \left(\left(\overset{T}{\underset{i=n-1,0}{\otimes}} \begin{bmatrix} x_i \textcircled{T}\, 0 \\ x_i \textcircled{T}\, 1 \\ e_{\textcircled{T}} \end{bmatrix} \right) \overset{T}{\otimes} \begin{bmatrix} y \textcircled{T}\, 0 \\ y \textcircled{T}\, 1 \\ e_{\textcircled{T}} \end{bmatrix} \right) .$$

□

2.3.4. Prime implicants and prime implicates

From the redundant expansions (2.44) we may obtain other classical canonical expansions using the disjunction and the conjunction. We shall consider the representations of a function f as the disjunction of all its *prime implicants* and as the conjunction of all its *prime implicates.*

We call *cube* any conjunction of letters $(x_i^{(e_i)} \wedge \ldots \wedge x_j^{(e_j)})$; we call *implicant* of a Boolean function any cube smaller than or equal to this function and *prime implicant* an implicant which is not contained in any other implicant of this function.

In view of theorem 2.3.3. the redundant disjunctive expansion of f is written in matrix form :

$$f(\underline{x}) = \underline{\phi}^{(\wedge)}(f)\ [\vee\wedge] \quad \left(\overset{\wedge}{\underset{i=n-1,0}{\otimes}} \begin{bmatrix} \bar{x}_i \\ x_i \\ 1 \end{bmatrix} \right) . \tag{2.45}$$

The 3^n elements of the vector $\underline{\phi}^{(\wedge)}(f)$ are in one-to-one relation with the 3^n subcubes of the n-cube ; in a more precise way to the 1 of $\underline{\phi}^{(\wedge)}(f)$ correspond constant subcubes with a value 1 while to the 0 of $\underline{\phi}^{(\wedge)}(f)$ correspond subcubes with at least one zero. Thus the expansion (2.45) gives us the function $f(\underline{x})$ as a disjunction of *all* its implicants. The representation of $f(\underline{x})$ as a disjunction of all its prime implicants is then obtained from (2.45) by dropping the implicants contained in other ones (application of the absorption law $a \vee b = a$ if $b < a$). We call *anticube* any disjunction of letters $(x_i^{(e_i)} \vee \ldots \vee x_j^{(e_j)})$; an *implicate* of a Boolean function is an anticube greater than or equal to this function and an implicate will be said to be a *prime implicate* if it is not larger than any other implicate of this function. Arguments dual to those developed above allow us to deduce, from the redundant conjunctive expansion of f :

$$f(\underline{x}) = \underline{\phi}^{(\vee)}(f)[\wedge \vee] \ (\overset{\vee}{\underset{i=n-1,0}{\otimes}} \begin{bmatrix} x_i \\ \bar{x}_i \\ 0 \end{bmatrix}) \ , \tag{2.46}$$

the canonical representation of f as the conjunction of all its prime implicates.

2.3.5. Galoisian expansions

Theorem

The 2^n Galoisian expansions of $f(\underline{x})$ *are obtained from the matrix relation :*

$$\underline{\phi}^{(\textcircled{T})}(f)\ [\textcircled{T}\ T]\ (\overset{T}{\underset{i=n-1,0}{\otimes}} \begin{bmatrix} 1 & 0 \\ 0 & 1 \\ x_i \textcircled{T} 0 & x_i \textcircled{T} 1 \end{bmatrix}) \ . \tag{2.47}$$

Proof

The proof derives from the same kind of recurrent relation than that used for theorem 2.3.3., the initial relation being in the present case (2.30). □

We may obtain an equivalent form for the matrix relation (2.45) by starting the recurrent scheme with the relation (2.27) instead of (2.30); we obtain :

$$f(\underline{x}) = \underline{\phi}^{(\textcircled{T})}(f)\ [\textcircled{T}\ T]\ (\overset{T}{\underset{i=n-1,0}{\otimes}} \begin{bmatrix} h_i \textcircled{T} 1 \\ h_i \textcircled{T} 0 \\ x_i \textcircled{T} h_i \end{bmatrix}) \ . \tag{2.48}$$

The 2^n expansions are obtained by giving to the vector $\underline{h}=(h_{n-1},\ldots,h_1,h_0)$ its 2^n possible values.

Remember also that if $\textcircled{T}$ means the modulo-2 sum $\oplus$, the expansions (2.47) are also called *Newton* expansions, *Taylor* expansions or *Reed-Muller* expansions.

3. Differences of Boolean functions

3.1. Simple differences

3.1.1. Definitions

Let A be a subset of $B_2=\{0,1\}$; A is thus either the empty set $\emptyset$, either $\{0\}$, either $\{1\}$, or the set $B_2=\{0,1\}$.

We define the *Boolean exponentiation* $x_T^{(A)}$ of x with respect to A, in the following way :

$$x_T^{(A)} = e_T \text{ if } x \in A , \qquad = e_\perp \text{ if } x \notin A .$$

We call *difference of f with respect to x* the function denoted $\uparrow^A f/\uparrow x$ and defined as follows :

$$\frac{\uparrow^A f}{\uparrow x} = f \uparrow f(x \circledT x_T^{(A)}) : \tag{3.1}$$

The notation $f(x \circledT x_T^{(A)})$ means that in the expansion of f one replaces x by $x \circledT x_T^{(A)}$. Since $e_T = \bar{e}_\perp$,we may write :

$$x \circledT x_T^{(A)} = x \circledperp x_\perp^{(A)} .$$

Since relation (3.1) does not change when substituting $\perp$ for T , we write (3.1) in the more usual form :

$$\frac{\uparrow^A f}{\uparrow x} = f \uparrow f(x \oplus x^{(A)}), \tag{3.1 bis}$$

where $x^{(A)}$ means $x_\wedge^{(A)}$.

According to the four possible values of A, we may write :

$$\frac{\uparrow^\emptyset f}{\uparrow x} = f \uparrow f ,$$

$$\frac{\uparrow^{\{0\}} f}{\uparrow x} = f \uparrow f(1) ,$$

$$\frac{\uparrow^{\{1\}} f}{\uparrow x} = f \uparrow f(0) ,$$

$$\frac{\uparrow^{\{0,1\}} f}{\uparrow x} = f \uparrow f(\bar{x}) .$$

When considering the complete set $B_2=\{0,1\}$ we shall replace the notation $\uparrow^{\{0,1\}}f/\uparrow x$ by $\uparrow f/\uparrow x$. We may express the difference with respect to $x^{(A)}$ (considered here as an auxiliary variable) according to the expansions studied in section 2.2.

Lagrange expansion :

$$\frac{\uparrow^{A} f}{\uparrow x} = [\,(x^{(A)} \,Ⓣ\, 0) \perp \frac{\uparrow^{\emptyset} f}{\uparrow x}]\ \mathrm{T}\ [\,(x^{(A)} \,Ⓣ\, 1) \perp \frac{\uparrow f}{\uparrow x}\,] \, ; \tag{3.2}$$

Redundant expansion :

$$\frac{\uparrow^{A} f}{\uparrow x} = [\,(x^{(A)} \,Ⓣ\, 0) \perp \frac{\uparrow^{\emptyset} f}{\uparrow x}]\ \mathrm{T}\ [\,(x^{(A)} \,Ⓣ\, 1) \perp \frac{\uparrow f}{\uparrow x}]\ \mathrm{T}\ [\ \frac{\uparrow^{\emptyset} f}{\uparrow x} \perp \frac{\uparrow f}{\uparrow x}] \ ; \tag{3.3}$$

Galoisian expansion :

$$\frac{\uparrow^{A} f}{\uparrow x} = [\,h\ \mathrm{T}\ \frac{\uparrow^{\emptyset} f}{\uparrow x}]\ Ⓣ\ [\,\bar{h}\ \mathrm{T}\ \frac{\uparrow f}{\uparrow x}]\ Ⓣ\ \ [\,(x^{(A)} Ⓣ\, h)\ \mathrm{T}\ (\frac{\uparrow^{\emptyset} f}{\uparrow x}\ Ⓣ\ \frac{\uparrow f}{\uparrow x})] \,. \tag{3.4}$$

3.1.2. Functional properties

Before particularizing the operation $\uparrow$ and the set A, we shall state some functional properties of $\uparrow^{A} f/\uparrow x$.

P.1. If the function f is a constant (that we denote c), we deduce from (3.1) :

$$\frac{\uparrow^{A} c}{\uparrow x} = c \uparrow c = c' \ . \tag{3.5}$$

Thus the difference $\uparrow^{A} c/\uparrow x$ of a constant is also a constant. The property P.1. remains true if the term *constant* is replaced by *function degenerated in x* (for a definition of the *degeneracy* see 3.1.8).

P.2. The difference $\uparrow f/\uparrow x$ is degenerated in x ; indeed :

$$\frac{\uparrow f}{\uparrow x} = f \uparrow f(\bar{x}) \ ,$$

$$= \bar{x}\,[\,f(0) \uparrow f(1)] \vee x[f(1) \uparrow f(0)] \ = \ f(0) \uparrow f(1) \ . \tag{3.6}$$

P.3. (*Linearity*) Be $\uparrow_d$ the law which distributes $\uparrow$ (see 2.3.2.) ; from (3.1) we deduce :

$$\begin{aligned} \uparrow^{A}(c \uparrow_d f)/\uparrow x &= (c \uparrow_d f) \uparrow (c \uparrow_d f(x \oplus x^{(A)})) \ , \\ &= c \uparrow_d (f \uparrow f(x \oplus x^{(A)})) \ , \\ &= c \uparrow_d (\uparrow^{A} f/\uparrow x) \ . \end{aligned} \tag{3.7}$$

For any pair of functions f_0, f_1, we have :

$$
\begin{aligned}
\uparrow^{A}(f_0\uparrow f_1)/\ x &= (f_0\uparrow f_1)\uparrow(f_0(x\oplus x^{(A)})\uparrow f_1(x\oplus x^{(A)}))\ ,\\
&= (f_0\uparrow f_0(x\oplus x^{(A)}))\uparrow(f_1\uparrow f_1(x\oplus x^{(A)}))\ ,\\
&= (\uparrow^{A} f_0/\uparrow x)\uparrow(\uparrow^{A} f_1/\uparrow x)\ .
\end{aligned}
\qquad (3.8)
$$

The relations (3.7) and (3.8) may be gathered in order to give the following linearity property :

$$
\frac{\uparrow^{A}[(c_0\uparrow_d f_0)\uparrow(c_1\uparrow_d f_1)]}{\uparrow x} = (c_0\uparrow_d\frac{\uparrow^{A} f_0}{\uparrow x})\uparrow(c_1\uparrow_d\frac{\uparrow^{A} f_1}{\uparrow x})\ . \qquad (3.9)
$$

P.4. *(Duality)*

The operations denoted $\uparrow$ and $\downarrow$ are said to be *dual* to each other if they satisfy the relation :

$$
a\uparrow b = \overline{\bar{a}\downarrow\bar{b}}\ . \qquad (3.10)
$$

We consider the following pairs of operations $(\uparrow,\downarrow)$:

$$
(\wedge,\vee)\ \text{and}\ (\oplus,\odot)\ .
$$

$$
\begin{aligned}
\frac{\uparrow^{A} f}{\uparrow x} &= f\uparrow f(x\oplus x^{(A)})\ ,\\
&= \overline{\bar{f}\downarrow\bar{f}(x\oplus x^{(A)})}\ ,\\
&= \overline{\frac{\downarrow^{A}\bar{f}}{\downarrow x}}\ .
\end{aligned}
\qquad (3.11)
$$

3.1.3. Relations between the differences

After stating the most important functional properties of the differences, we present a first theorem which gives us the relations between the differences.

Theorem

(a)
$$
\frac{T^{A} f}{T x} = \frac{\odot^{A} f}{\odot x}\odot\frac{\perp^{A} f}{\perp x}\ ; \qquad (3.12)
$$

(b)
$$
\frac{\odot^{A} f}{\odot x} = \frac{\perp^{A} f}{\perp x}\ T\ \overline{\frac{T^{A} f}{T x}}\ . \qquad (3.13)
$$

Proof

The proof of (3.12, 3.13) results from the Stone relations (2.13,2.14) and from the duality property (3.11). By putting $a=f$ and $b=f(x\oplus x^{(A)})$ in the relations (2.13) and (2.14), the definition of the difference and the duality relation allow us to state the properties (3.12) and (3.13). □

We give an interpretation of theorems 3.1.3. and 3.1.4. (below) in terms of *switching theory* in section 3.2.8.

3.1.4. Relations between f and its differences

The following theorem states relations between the differences and the function f.

Theorem

(a) $$f \; T \; \frac{\textcircled{$\perp$}^A f}{\textcircled{$\perp$} x} = \frac{T^A f}{T x} \quad ; \tag{3.14}$$

(b) $$f \; T \; \frac{\textcircled{T}^A f}{\textcircled{T} x} = f \; \textcircled{T} \; \frac{T^A f}{T x} \quad ; \tag{3.15}$$

(c) $$f = \frac{T^A f}{T x} \perp \frac{T^{\bar{A}} f}{T x} \quad , \tag{3.16}$$

where $\bar{A}$ means the complement of A with respect to $\{0,1\}$

Proof

Consider the following two relations, the first one deriving from the Stone relation (2.13) and the second one being a consequence of the axioms (B_5) and (B_7) and of 2.1.3 :

$$a \; T \; (a \; \textcircled{$\perp$} \; b) = a \; T \; b \quad , \tag{3.17}$$

$$a \; T \; (a \; \textcircled{T} \; b) = a \; \textcircled{T} \; (a \; T \; b) \quad . \tag{3.18}$$

The formulas (3.14) and (3.15) derive from (3.17) and (3.18) respectively by putting $a=f$ and $b=f(x \oplus x^{(A)})$.

Let us now prove (3.16).

$$\begin{aligned} \frac{T^A f}{T x} \perp \frac{T^{\bar{A}} f}{T x} &= (f \; T \; f(x \oplus x^{(A)})) \perp (f \; T \; f(x \oplus x^{(\bar{A})})) \\ &= (f \perp f) \; T \; (f \perp f(x \oplus x^{(A)})) \; T \; (f \perp f(x \oplus x^{(\bar{A})})) \; T \\ &\qquad (f(x \oplus x^{(A)}) \perp f(x \oplus x^{(\bar{A})})) \quad \text{(by distributivity)} \\ &= f \; . \qquad \text{(by absorption)} \end{aligned}$$

□

3.1.5. Relations between differences for a given operation $\dagger$.

Theorems 3.1.3. and 3.1.4. state properties relating the differences associated to the various operations T and $\textcircled{T}$; the next theorem studies the properties relating the differences for a fixed operation $\dagger$ and for various values of A.

<u>Theorem</u>

$$\mathop{\textcircled{T}}_{A \subseteq \{0,1\}} \frac{\uparrow^{A} f}{\uparrow x} = e_{\textcircled{T}} \, . \tag{3.19}$$

<u>Proof</u>

The expansion (2.23) written in the form :

$$f(x) = ((x \textcircled{T} 0) \; T \; f(1)) \textcircled{T} ((x \textcircled{T} 1) \; T \; f(0)) \; ,$$

allows us to expand the function $\mathop{\textcircled{T}}_{A} (\uparrow^{A} f/\uparrow x)$, considered as a function of $x^{(A)}$, according to the scheme :

$$\begin{aligned}
\mathop{\textcircled{T}}_{A \subseteq \{0,1\}} \frac{\uparrow^{A} f}{\uparrow x} &= (f \uparrow f) \textcircled{T} (f \uparrow f(0)) \textcircled{T} (f \uparrow f(1)) \textcircled{T} (f \uparrow f(\bar{x})) \\
&= [(x \textcircled{T} 0) \, T ((f(1) \uparrow f(1)) \textcircled{T} (f(1) \uparrow f(0)) \textcircled{T} (f(1) \uparrow f(1)) \textcircled{T} (f(1) \uparrow f(0)))] \\
&\quad \textcircled{T} [(x \textcircled{T} 1) \, T ((f(0) \uparrow f(0)) \textcircled{T} (f(0) \uparrow f(0)) \textcircled{T} (f(0) \uparrow f(1)) \textcircled{T} (f(0) \uparrow f(1)))] \\
&= [(x \textcircled{T} 0) \, T \, e_{\textcircled{T}}] \textcircled{T} [(x \textcircled{T} 1) \, T \, e_{\textcircled{T}}] \quad \text{(in view of (2.17,2.18))} \\
&= e_{\textcircled{T}} \, .
\end{aligned}$$

□

3.1.6. <u>Table of differences</u>

Theorems 3.1.3-3.1.5. constitute the essential of the relations between the differences ; their use will appear progressively along the next sections. We consider mainly for $\uparrow$ the operations $\wedge, \vee$ and $\oplus$ while we take for A the set of the non-empty subsets of $\{0,1\}$. The most usual notations and callings for the differences are gathered in table 3.I.

From relations (3.12) and (3.19) we deduce that the modulo-2 sum of the elements of a row (or of a column if the row corresponding to $A=\emptyset$ is added) is identically zero. We may deduce a set of equalities relating the various differences by adding modulo-2 an arbitrary subset of rows and columns of the table 3.I augmented with the row $A=\emptyset$. Consider e.g. the functions $f \oplus qf/qx$ and $f \oplus pf/px$; these functions are related to the concept of oriented difference (see section 3.1.11). If we want to express these functions in terms of envelopes we must take the modulo-2 sum of the second and of the third column of table 3.I ; we obtain the following relations:

$$f \oplus \frac{qf}{qx} = \frac{q^{\{0\}} f}{qx} \oplus \frac{q^{\{1\}} f}{qx} \, ,$$

$$f \oplus \frac{pf}{px} = \frac{p^{\{0\}} f}{px} \oplus \frac{p^{\{1\}} f}{px} \, .$$

$\uparrow$	$\oplus$	$\wedge$	$\vee$
A	Boolean differences : $\frac{\Delta^{A} f}{\Delta x}$	Meet differences : $\frac{p^{A} f}{p x}$	Join differences : $\frac{q^{A} f}{q x}$
{0}	$\frac{\Delta^{\{0\}} f}{\Delta x} = f(1) \oplus f$ $= \bar{x}(f(0) \oplus f(1))$ Decreasing divided difference	$\frac{p^{\{0\}} f}{p x} = f(1) \wedge f$ $= f(1)(f(0) \vee x)$ Increasing lower envelope	$\frac{q^{\{0\}} f}{q x} = f(1) \vee f$ $= f(1) \vee \bar{x}\, f(0)$ Decreasing upper envelope
{1}	$\frac{\Delta^{\{1\}} f}{\Delta x} = f(0) \oplus f$ $= x(f(0) \oplus f(1))$ Increasing divided difference	$\frac{p^{\{1\}} f}{p x} = f(0) \wedge f$ $= f(0)(f(1) \vee \bar{x})$ Decreasing lower envelope	$\frac{q^{\{1\}} f}{q x} = f(0) \vee f$ $= f(0) \vee x\, f(1)$ Increasing upper envelope
{0,1}	$\frac{\Delta f}{\Delta x} = f(0) \oplus f(1)$ Boolean difference	$\frac{p f}{p x} = f(0) \wedge f(1)$ Meet difference	$\frac{q f}{q x} = f(0) \vee f(1)$ Join difference

Table 3.I.

If finally we take the modulo-2 sum of the rows corresponding to $A=\emptyset$ and $A=\{0,1\}$ and of the third column, we obtain :

$$f \oplus \frac{pf}{px} = \frac{q^{\{1\}}f}{qx} \oplus \frac{q^{\{0\}}f}{qx} \oplus \frac{\Delta f}{\Delta x} .$$

3.1.7. Lattice of differences

Taking into account the property (2.3) and the axiom (A_3) of section 2.1.2, we have :

$$\frac{T^{\emptyset}f}{Tx} = f \mathrm{T} f = f ,$$

$$f \perp \frac{Tf}{Tx} = f \perp (f \mathrm{T} f(\bar{x})) = f .$$

These two equalities allow us to write the redundant expansion (3.3) of $T^A f/Tx$ in the following form :

$$\frac{T^A f}{Tx} = f \mathrm{T} [(x^{(A)} ⦶ 1) \perp \frac{Tf}{Tx}] . \tag{3.20}$$

By giving successively to T the meanings of disjunction and of conjunction, we deduce from (3.20) :

$$\frac{q^A f}{qx} = f \vee x^{(A)} \frac{qf}{qx} , \tag{3.21}$$

$$\frac{p^A f}{px} = f(x^{(\bar{A})} \vee \frac{pf}{px}) . \tag{3.22}$$

Be A_0 and A_1 two values of A ; f(A) is *increasing* in A if $A_0 \geqslant A_1$ implies $f(A_0) \geqslant f(A_1)$. If $A_0 \geqslant A_1$ implies $f(A_0) \leqslant f(A_1)$, f(A) is *decreasing* in A. A function which is at the same time increasing and decreasing in A is *degenerated* in A, i.e. $f(A_0)=f(A_1)$ $\forall A_0,A_1$.

The function $x^{(A)}$ is increasing in A ; the lattice of values of $x^{(A)}$ is given as example in figure 3.1. From relations (3.21) and (3.22) we deduce that $q^A f/qx$ is an increasing function in A while $p^A f/px$ is a decreasing function in A. From the inequalities :

$$a \vee b \geqslant a \geqslant ab \text{ and } a \vee b \geqslant a \oplus b ,$$

we deduce :

$$\frac{q^A f}{qx} \geqslant f \geqslant \frac{p^A f}{px} \quad \text{and} \quad \frac{q^A f}{qx} \geqslant \frac{\Delta^A f}{\Delta x} . \tag{3.23}$$

From the inequalities (3.23), from the increasing and decreasing character of the functions $q^A f/qx$ and $p^A f/px$ respectively and from their expressions for the various values of A (see e.g. table 3.I or (3.21,3.22)) we deduce that the set of differences $T^A f/Tx$ is a distributive lattice.

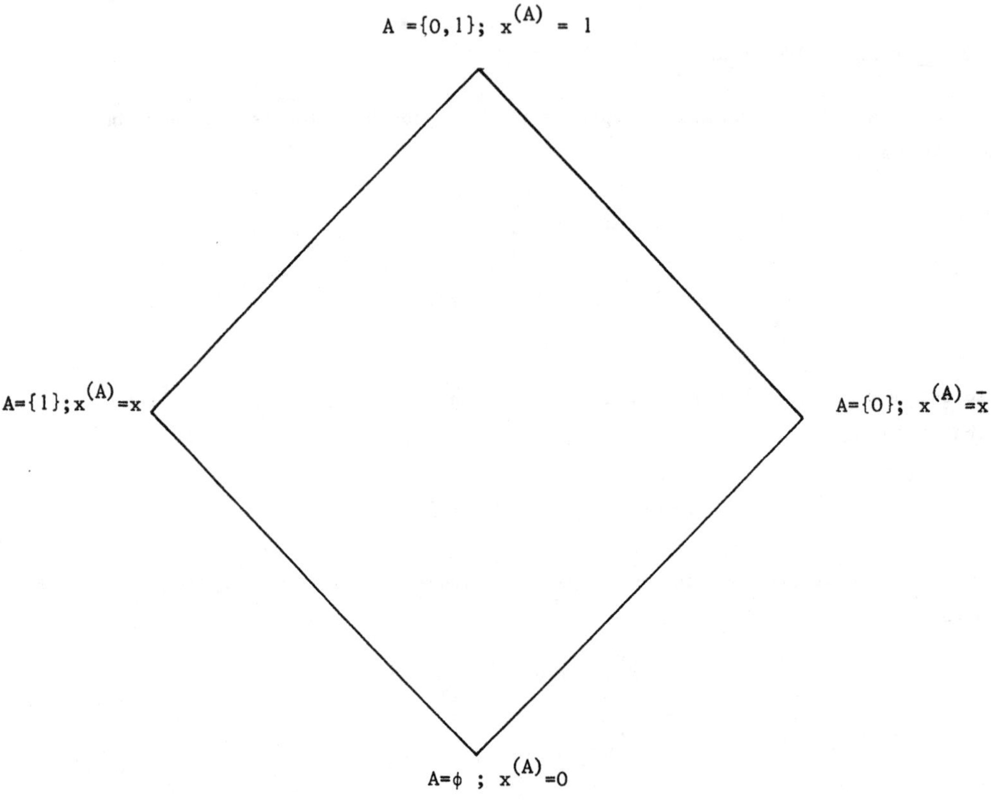

Figure 3.1. Lattice of the values of $x^{(A)}$.

Consider the difference operators $\textcircled{T}^A f/\textcircled{T}x$; from the Lagrange expansion (3.2) we deduce :

$$\frac{\textcircled{T}^A f}{\textcircled{T}x} = (x^{(A)} \textcircled{T}\, 0)\ T\ \frac{\textcircled{T}f}{\textcircled{T}x} \quad . \tag{3.24}$$

If we consider the difference operator associated to the identity, i.e. :

$$\frac{\Gamma^A f}{\Gamma x} = f \odot f(x \oplus x^{(A)}) = \overline{\frac{\Delta^A f}{\Delta x}} \ , \tag{3.25}$$

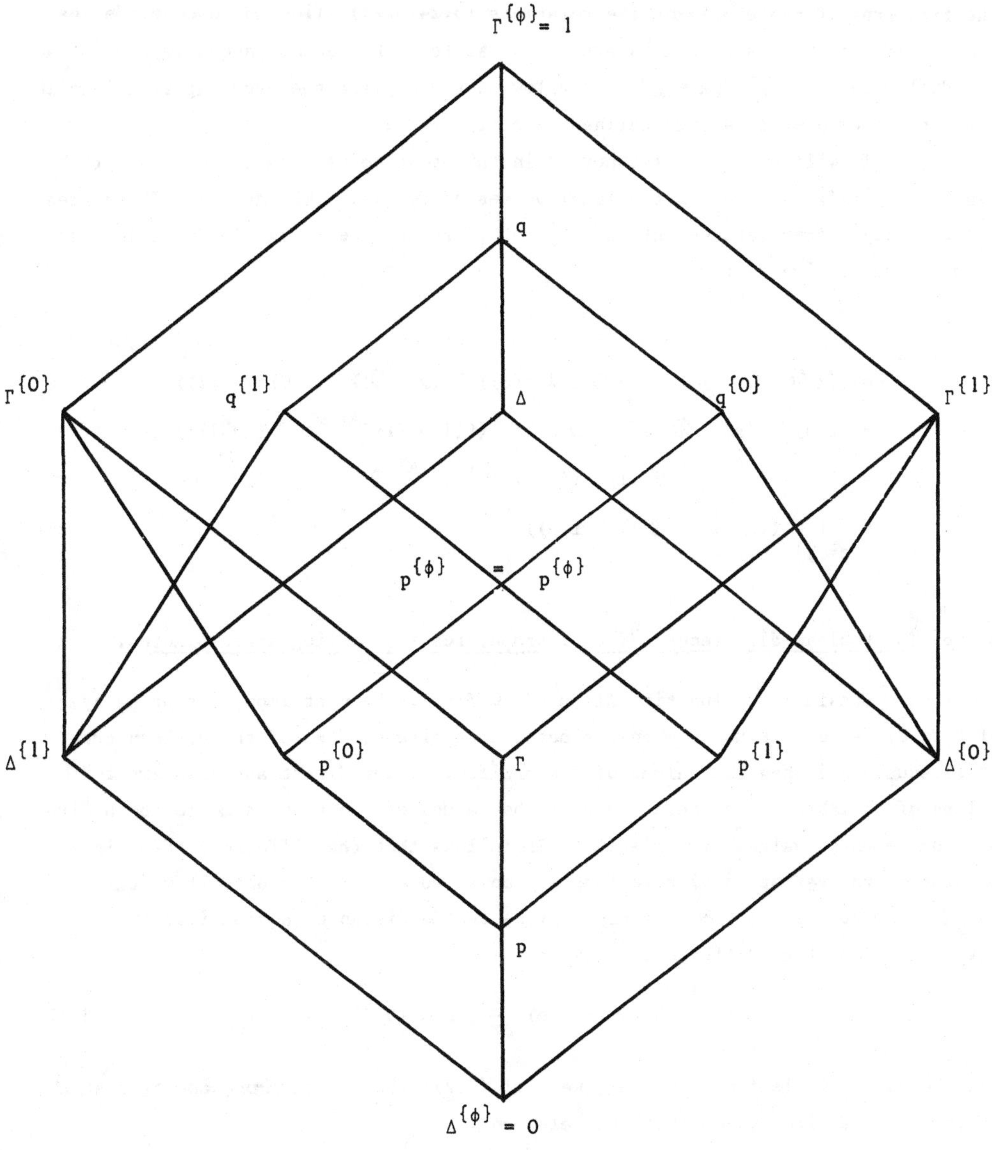

Figure 3.2. Lattice of the difference operators

the first row of table 3.I and the relations (3.24,3.25) allow us to conclude that the set of differences $\textcircled{T}^A f/\textcircled{T}x$ is a Boolean lattice. The second inequality (3.23) and its dual relation : $p^A f/px \leqslant \Gamma^A f/\Gamma x$ allow us to complete the ordering relations in the lattice of operators thus defined (see figure 3.2).

It will be useful further on in this text to have an expression for the expansion $T^A f/Tx$ in terms of the local values of f, i.e. f(0) and f(1). The expressions (2.21) (after replacement of $T,\perp$ by $\perp,T$ respectively) and (3.20) allow us to write successively :

$$\begin{aligned}\frac{T^A f}{Tx} &= ((x \textcircled{T} 0)\perp f(0))\ T\ ((x \textcircled{T} 1)\perp f(1))\ T\ ((x^{(A)} \textcircled{T} 1)\perp (f(0)\ T\ f(1)))\ ,\\ &= (f(0)\perp ((x^{(A)} \textcircled{T} 1)\ T\ (x \textcircled{T} 0)))\ T\ (f(1)\perp ((x^{(A)} \textcircled{T} 1)\ T\ (x \textcircled{T} 1)))\ ,\\ &= (f(0)\perp (x^{(0\cup A)} \textcircled{T} 1))\ T\ (f(1)\perp (x^{(1\cup A)} \textcircled{T} 1))\ ,\\ &= \underset{e=0,1}{T}\ ((x^{(e\cup A)} \textcircled{T} 1)\perp f(e))\ .\end{aligned} \tag{3.26}$$

3.1.8. The Boolean difference $\Delta^A f/\Delta x$: properties and bibliographical notes.

Consider the function $\Delta f/\Delta x = f \oplus f(\bar{x})$; the most important property of this difference is that, when evaluated at a given vertex of the Boolean cube, it is equal to 1 when the values of f are different for direct and complemented values of variable x, and equal to 0 if the values of f are the same for both direct and complemented values of variable x. It follows that the difference $\Delta f/\Delta x$ is a measure of the variation of this function on an edge of the n-cube. If we take for $\textcircled{T}$ the modulo-2 sum : $\oplus$, we may express the Galoisian expansion (2.27) in terms of the Boolean difference, i.e. :

$$f = f(h) \oplus (x \oplus h)\ \frac{\Delta f}{\Delta x}\ ,\ h \in \{0,1\}\ . \tag{3.27}$$

For the one-variable function f(x), we call (3.27) the *Newton expansion* of f at the vertex h. Note already now that the relation :

$$f \oplus f(x \oplus x^{(A)}) = x^{(A)}\ \frac{\Delta f}{\Delta x}\ , \tag{3.28}$$

which we deduce from (3.24), will be used to deal with the *fault detection theory*. The *Boolean difference* $\Delta f/\Delta x$, introduced by Reed [1954] , has been more thoroughly investigated in a paper by Akers [1959] who obtained the algebraic properties of the Boolean difference. With the help of Boolean difference, Akers was able to determine the conditions under which changes in the variables of a switching function would cause change in the function. This led to the discovery of a number of relationships and theorems for switching functions that correspond to those of finite differences and of differential calculus of real functions (Akers [1959], Thayse [1971], Thayse and Davio [1973], Rudeanu [1974], pp. 328-338, Lee [1978], chapter 3). For example, Akers gives a series expansion for switching functions that closely resembles the Taylor-MacLaurin series. Amar and Condulmari [1967] and Sellers, Hsiao and Bearnson [1968] applied the Boolean differences $\Delta^{\{0\}}f/\Delta x$ and $\Delta^{\{1\}}f/\Delta x$ to the problem of fault diagnosis ; this gave rise to a considerable number of papers using the Boolean difference for fault detection (Davio and Piret [1969], Yau and Tang [1971]). Thayse [1971] and Thayse and Davio [1973] describe a *Boolean differential calculus* which encompases and generalizes the algebraic concepts introduced by the authors quoted above ; it is considered as a general method of representation of an arbitrary switching function and as such could be applied to any classical switching problem. Moreover Thayse, Pichat and Lapscher have shown that the Boolean difference is applicable to a number of areas other than fault diagnosis, e.g. hazard detection, function decomposition and analysis and synthesis of sequential switching networks (Akers [1959], Pichat [1968], Shen, Mc Kellar and Weiner [1971], Thayse [1972c], Lapscher [1972], Bioul and Davio [1972]).

3.1.9. The meet difference $p^A f/px$: properties and bibliographical notes

Let us consider the meet difference $pf/px=f(1)f(0)$; this function is degenerate in x, it is smaller than or equal to f and is equal to f in at least one of the vertices x=0 or x=1. It is thus the greatest function degenerate in x and smaller than or equal to f. Consider now the function : *disjunction of the prime implicants of f degenerate in x*. Any function f being the disjunction of *all* its prime implicants, it is thus a function smaller than or equal to f ; it is the greatest function smaller than or equal to f and degenerate in x since if there was a greater function, at least one of its prime implicants would be greater and would be moreover a prime implicant of f. This is impossible, since we have taken all the prime implicants of f degenerate in x. We may state the following proposition.

Proposition

The meet difference pf/px is a function equal to the disjunction of all the prime implicants of f degenerate in x.

We shall now introduce a more general concept of degeneracy and relate this concept to the function $p^A f/px$.

We say that a function f is *A-degenerate* in x if and only if it satisfies the relation :

$$1^{(A)}\bar{f}(0)f(1) \vee 0^{(A)}f(0)\bar{f}(1)=0. \tag{3.29}$$

If A is the set {0,1} , the *{0,1}-degeneracy* coincides with the *degeneracy* defined in 3.1.7 ; indeed, in this case $1^{(\{0,1\})}=0^{(\{0,1\})}=1$ and the relation (3.29) is verified if and only if f(0)=f(1). If A is the element {0} , the *{0}-degeneracy* coincides

with the concept of *increasing function* defined in 3.1.6., indeed, since $1^{(\{0\})}=0$ and $0^{(\{0\})}=1$, the relation (3.29) is verified if and only if $f(0)\bar{f}(1)=0$, i.e. if and only if f is increasing in x. An *increasing function* is thus a *{0}-degenerate function*. We verify that a *decreasing function* is a *{1}-degenerate function*.
Mc Naughton [1961] has shown the following proposition (expressed here in terms of *{h}-degeneracy*, $h \in \{0,1\}$) :

Proposition

A function is {h}-degenerate in x if and only if it may be written in terms of a well-formed expression (see section 2.1.5) where $x^{(h)}$ *does not appear.*

It is clear that if a function f may be written as a well-formed expression without using the variable x, either under direct or under complemented form, this function is degenerate in x. The concept of A-degeneracy may be written in a simple way in terms of a well-formed expression.

Theorem

The function $p^A f/px$ *is a function A-degenerate in x equal to the disjunction of the prime implicants of f A-degenerate in x ; it is the greatest function A-degenerate in x and smaller than f.*

Proof

For A={0,1} ,we have shown that pf/px is a function degenerate in x and equal to the disjunction of the prime implicants of f degenerate in x ; it is the greatest function degenerate in x and smaller than or equal to f. For A= {h} , we have :

$$\frac{p^{\{h\}}f}{px} = x^{(\bar{h})} f(\bar{h}) \vee f(0)f(1) . \tag{3.30}$$

This function is h-degenerate in x, smaller than or equal to f and it is the disjunction of the prime implicants of f {h}-degenerate in x. □

Corollary

A function f is A-degenerate in x if and only if $f=p^A f/px$.

The meet difference pf/px was defined by the author (Thayse [1973] , [1976]) who related it to the classical concept of implicant of a function. This difference has been used as a theoretical support for designing several types of algorithms detecting the prime implicants of a function (Thayse [1973a] , [1976], [1978] , Deschamps and Thayse [1973a] , Davio, Deschamps and Thayse [1978]).
The meet difference has also been used by Davio and Thayse [1973] for the analysis of fuzzy functions and by Thayse [1978] for a new formulation of the consensus theory.

The *lower envelopes* (which are nothing but the meet differences $p^{\{h\}}f/px$ - see Table 3.I) were defined and studied by Kuntzmann [1965] . Their properties were used by Deschamps and the author (Thayse [1976] , Thayse and Deschamps [1977]) for characterizing increasing and decreasing functions and for the synthesis of logical networks.

Note also that the meet and join differences are dual concepts ; so are also the concepts of implicant and implicate and of lower envelope and upper envelope. For reasons of symmetry and of duality we shall state the remaining properties and interpretations of the meet differences after the join differences have been studied.

3.1.10. The join difference $q^A f/qx$: properties and bibliographical notes

The join difference $qf/qx=f(1) \vee f(0)$ is a function degenerate in x ; it is greater than or equal to f and reaches the value of f at least at one vertex. It is thus the smallest function degenerate in x and greater than or equal to f.

Arguments dual to those developed for the meet difference allow us to state the following proposition.

Proposition

The function qf/qx is a function degenerate in x equal to the conjunction of the prime implicates of f degenerate in x.

Any (prime) implicate (resp. implicant) of a function f is the complement of a (prime) implicant (resp. implicate) of $\bar{f}$. Indeed, the complement of a cube is an anticube and conversely. Moreover if m is a prime implicant of f and m' an implicant :

$$f \geqslant m \geqslant m' \Leftrightarrow \bar{f} \leqslant \bar{m} \leqslant \bar{m}' \quad .$$

Thus the function f has as many (prime) implicants as $\bar{f}$ has (prime) implicates. Relation (3.11) allows us to write :

$$\overline{\frac{q^A f}{qx}} = \frac{p^A \bar{f}}{px} \quad . \tag{3.31}$$

Theorem 3.1.8 and the preceding arguments allow us to state the following theorem.

Theorem

The function $\overline{q^A f/qx}$ is a function A-degenerate in x equal to the disjunction of the prime implicants of $\bar{f}$ A-degenerate in x ; it is the greatest function A-degenerate in x and smaller than or equal to $\bar{f}$.

The above theorem is stated in terms of the function $\bar{f}$ instead of f ; a translation of this theorem in terms of f leads to the two statements below.

1. *The function qf/qx is a function degenerate in x, equal to the conjunction of the prime implicates of f degenerate in x ; it is the smallest function degenerate in x and greater than or equal to f.*

2. *The function* $q^{\{h\}}f/qx$, $h \in \{0,1\}$, *is a function* $\{\bar{h}\}$*- degenerate in* x, *equal to the conjunction of the prime implicates of* f $\{\bar{h}\}$*-degenerate in* x *; it is the smallest function* $\{\bar{h}\}$*-degenerate in* x *and greater than or equal to* f.

The join difference was defined by Fadini [1961] and used by Lapscher [1972] for the decomposition of functions. The author (Thayse [1973a], [1976]) related the concept of join difference to that of implicate of a Boolean function and obtained algorithms for the obtention of the prime implicates. Kuntzmann [1965] defined the *upper envelopes* (or join differences $q^{\{h\}}f/qx$ - see Table 3.I)) while the author (Thayse [1976], Thayse and Deschamps [1977]) used these notions for the characterization of increasing and of decreasing functions.

3.1.11. Properties of the envelopes

The concepts of *lower* and of *upper envelopes* were introduced by Kuntzmann [1965]; these concepts are useful in the characterization of functional properties of Boolean functions. If, e.g. we give successively to T the meanings $\vee$ and $\wedge$ and if we take for A the set $\{0\}$, the relation (3.16) will be stated as follows :

-*The function* f *is the conjunction of its increasing and decreasing upper envelopes* ;

-*The function* f *is the disjunction of its increasing and decreasing lower envelopes* .

It is clear that for an increasing (resp. decreasing) function f its increasing (resp. decreasing) upper and lower envelopes are equal to f.

Theorem

The truth of one of the three relations :

$$f \oplus \frac{p^{A}f}{px} \equiv 0 \ , \qquad (3.32)$$

$$\bar{f} \oplus \frac{q^{\bar{A}}\bar{f}}{qx} \equiv 0 \ , \qquad (3.33)$$

$$\frac{p^{A}f}{px} \oplus \frac{q^{\bar{A}}f}{qx} \equiv 0 \ , \qquad (3.34)$$

is a necessary and sufficient condition for f *to be A-degenerate in* x $(A \subseteq \{0,1\}\ ,\ A \neq \emptyset)$.

Proof

Let A be the set $\{0,1\}$;

$$f \oplus \frac{pf}{px} = \bar{x}f(0)\ \bar{f}(1) \vee x\bar{f}(0)f(1) \ .$$

If f is degenerate in x, $f(0)=f(1)$ and $f \oplus pf/px=0$.

Conversely, if $f \oplus \frac{pf}{px} = 0$ for any value of x, then $\bar{f}(0)f(1)$ and $f(0)\bar{f}(1)$ must simultaneously be equal to zero ; this may happen only if $f(0)=f(1)$. The condition (3.32) thus holds for $A=\{0,1\}$. For A, either equal to $\{0\}$, or equal to $\{1\}$, the condition

(3.32) means that a function is increasing (resp. decreasing) in x if and only if it coincides with its increasing (resp. decreasing) envelope.
Relation (3.33) is obtained from (3.32) by simultaneous complementation of its two sides.
Relation (3.34) coincides with (3.32) when A is {0,1}; for A being either {0} or {1} the relation (3.34) means that a function is increasing (resp. decreasing) in x if and only if its increasing (resp. decreasing) upper and lower envelopes coincide. □

We shall compute several expressions for the function $f \oplus p^A f/px$; these expressions will be used in section 3.1.12.

$$f \oplus \frac{p^A f}{px} = f\frac{\Delta^A f}{\Delta x} \quad \text{(from 3.15))} \quad , \tag{3.35a}$$

$$= x^{(A)}(f \oplus \frac{pf}{px}) \quad \text{(from 3.22)} \quad , \tag{3.35b}$$

$$= x^{(1)}(x\,\bar{f}(0)f(1) \oplus \bar{x}\,f(0)\,\bar{f}(1)) \; , \tag{3.35c}$$

$$= x^{(1 \cap A)}\bar{f}(0)f(1) \oplus x^{(0 \cap A)}f(0)\bar{f}(1) \; . \tag{3.35d}$$

We verify that the Boolean exponentiation satisfies the following relations :

$$x^{(A)} \wedge x^{(B)} = x^{(A \cap B)} \quad , \tag{3.36a}$$

$$x^{(A)} \vee x^{(B)} = x^{(A \cup B)} \quad , \tag{3.36b}$$

$$x^{(A)} \oplus x^{(B)} = x^{(A \nabla B)} \quad .$$

where "∇" means the set theoretic symmetric difference. Consider the function $f \oplus q^A f/qx$; it may take the following forms :

$$f \oplus \frac{q^A f}{qx} = \bar{f}\,\frac{\Delta^A f}{\Delta x} \quad , \tag{3.37a}$$

$$= x^{(A)}(f \oplus \frac{qf}{qx}) \quad , \tag{3.37b}$$

$$= x^{(A)}(\bar{x}\,\bar{f}(0)f(1) \oplus x\,f(0)\bar{f}(1)), \tag{3.37c}$$

$$= x^{(0 \cap A)}\bar{f}(0)f(1) \oplus x^{(1 \cap A)}f(0)\bar{f}(1) \; . \tag{3.37d}$$

3.1.12. Envelopes and Galoisian expansions

The following theorem allows us to obtain the expression of the envelopes of f from the Galoisian expansions of this function.

First of all consider the expansion (2.27), i.e. :

$$f = f(h) \textcircled{\top} [(x \textcircled{\top} h) \top \frac{\textcircled{\top} f}{\textcircled{\top} x}] . \tag{3.38}$$

Observe that in the expression (3.38), $(x \textcircled{\top} h)$ will be written x or $\bar{x}$ according to the value of h and that the operation $\textcircled{\top}$ does not directly appear in this relation ; the same remark holds true for the difference $\textcircled{\top} f/\textcircled{\top} x$.

<u>Theorem</u>

The envelopes of f with respect to x are obtained from the Galoisian expansions with respect to x by substituting the operation $\top$ *for* $\textcircled{\perp}$.

<u>Proof</u>

By definition, we have :

$$\frac{\perp^{\{\bar{h}\}} f}{\perp x} = f(h) \perp f . \tag{3.39}$$

Replacing f by its Galoisian expansion (3.38) and taking into account the formula $a \top (a \textcircled{\perp} b) = a \top b$, we successively obtain :

$$\begin{aligned} \frac{\perp^{\{\bar{h}\}} f}{\perp x} &= f(h) \perp \{f(h) \textcircled{\top} [(x \textcircled{\top} h) \top \frac{\textcircled{\top} f}{\textcircled{\top} x}] \}, \\ &= f(h) \perp [(x \textcircled{\top} h) \top \frac{\textcircled{\top} f}{\textcircled{\top} x}] . \qquad \square \end{aligned} \tag{3.40}$$

It could also be possible to start from the Galoisian expression (2.48), i.e. :

$$f = f(h \top f(0)) \textcircled{\top} (\bar{h} \top f(1)) \textcircled{\top} (x^{(h)} \top \frac{\textcircled{\top} f}{\textcircled{\top} x}) ,$$

and then obtain the following expression for the envelopes :

$$\frac{\perp^{\{h\}} f}{\perp x} = (h \top f(0)) \perp (\bar{h} \top f(1)) \perp (x^{(h)} \top \frac{\textcircled{\top} f}{\textcircled{\top} x}) .$$

3.1.13. <u>The oriented differences : properties and bibliographical notes.</u>

It is useful to mention the definitions of the oriented differences. These operators are currently used to state several properties of the Boolean functions. The definitions are the following :

$$\begin{aligned} &\textit{increasing difference :} \quad \frac{\Delta f}{\Delta^{1} x} = \bar{f}(0) f(1) , \\ &\textit{decreasing difference :} \quad \frac{\Delta f}{\Delta^{0} x} = f(0) \bar{f}(1) . \end{aligned} \tag{3.41}$$

These functions are related to the increasing or decreasing character of f in x; we have :

$\Delta f/\Delta^{1} x = 0$ if and only if f is decreasing in x ,

$$\Delta f/\Delta^0 x = 0 \text{ if and only if f is increasing in x ,}$$

and :

$$\frac{\Delta f}{\Delta x} = \frac{\Delta f}{\Delta^0 x} \oplus \frac{\Delta f}{\Delta^1 x} \quad . \tag{3.42}$$

Relation (3.35c) allows us to write :

$$f \oplus \frac{pf}{px} = x \frac{\Delta f}{\Delta^1 x} \oplus \bar{x} \frac{\Delta f}{\Delta^0 x} \quad . \tag{3.43}$$

We have also :

$$f \oplus \frac{qf}{qx} = x \frac{\Delta f}{\Delta^0 x} \oplus \bar{x} \frac{\Delta f}{\Delta^1 x} \quad . \tag{3.44}$$

<u>Theorem</u>

The function $x^{(h)}(\Delta f/\Delta^h x)$ *is the disjunction of the prime implicants of* f $\{\bar{h}\}$*-degenerate in* x *but not* $\{0,1\}$ *-degenerate in* x.

<u>Proof</u>

In view of theorem 3.1.8, the disjunction of the prime implicants of f h -degenerate in x and not $\{0,1\}$-degenerate in x is :

$$\overline{\frac{pf}{px}} \; \frac{p^{\{h\}}f}{px} = (\bar{f}(0) \vee \bar{f}(1)) \; f(\bar{h})(f(h) \vee x^{(\bar{h})}) \; ,$$

$$= h(\bar{f}(0) \vee \bar{f}(1)) \; f(0)(f(1) \vee \bar{x}) \vee$$

$$\bar{h}(\bar{f}(0) \vee \bar{f}(1)) f(1)(f(0) \vee x) \; ,$$

$$= h \; \bar{x} \; f(0) \; \bar{f}(1) \vee \bar{h} \; x \; f(1) \; \bar{f}(0) \quad . \qquad \square$$

The oriented differences were defined by Talantsev [1959] ; they were meanly studied by East-European authors (see e.g. the papers by Lazarev and Piil [1961] ,[1962] ,[1963]). Brown and Young ([1969] ,[1970]) apply the oriented differences to the detection of stuck-faults. Tucker [1974] develops the oriented differences in the frame of a transition calculus for Boolean functions ; he defines also the oriented integrals which are nothing but the functions $f \oplus qf/qx = \bar{f}(\Delta f/\Delta x)$ and $f \oplus pf/px = f(\Delta f/\Delta x)$.

3.1.14. <u>Example</u>

$$f = x_0 x_2 \vee x_1 \bar{x}_2$$

$$\frac{\Delta f}{\Delta x_0} = x_2 \quad ; \quad \frac{\Delta f}{\Delta x_1} = \bar{x}_2 \quad ; \quad \frac{\Delta f}{\Delta x_2} = x_0 \oplus x_1 \quad ;$$

$$\frac{pf}{px_0} = x_1 \bar{x}_2 \quad ; \quad \frac{pf}{px_1} = x_0 x_2 \quad ; \quad \frac{pf}{px_2} = x_0 x_1 \quad ;$$

$$\frac{qf}{qx_0} = x_1 \vee x_2 \quad ; \quad \frac{qf}{qx_1} = x_0 \vee \bar{x}_2 \quad ; \quad \frac{qf}{qx_2} = x_0 \vee x_1 \quad ;$$

$$\frac{p^{\{0\}}f}{px_0} = x_0x_2 \vee x_1\bar{x}_2 \quad ; \quad \frac{p^{\{1\}}f}{px_0} = x_1\bar{x}_2 \quad ; \quad \frac{q^{\{0\}}f}{qx_0} = x_1 \vee x_2 \; ; \quad \frac{q^{\{1\}}f}{qx_0} = x_0x_2 \vee x_1\bar{x}_2 \; ;$$

$$\frac{p^{\{0\}}f}{px_1} = x_0x_2 \vee x_1\bar{x}_2 \quad ; \quad \frac{p^{\{1\}}f}{px_1} = x_0x_2 \quad ; \quad \frac{q^{\{0\}}f}{qx_1} = x_0 \vee \bar{x}_2 \quad ; \quad \frac{q^{\{1\}}f}{qx_1} = x_0x_2 \vee x_1\bar{x}_2 \quad ;$$

$$\frac{p^{\{0\}}f}{px_2} = x_0(x_1 \vee x_2) \quad ; \quad \frac{p^{\{1\}}f}{px_2} = x_1(x_0 \vee \bar{x}_2); \quad \frac{q^{\{0\}}f}{qx_2} = x_0 \vee x_1\bar{x}_2 \; ; \quad \frac{q^{\{1\}}f}{qx_2} = x_1 \vee x_0x_2 \quad .$$

These basic operators allow us e.g. to verify any one of the theorems and properties quoted in this chapter.

3.2. Multiple differences

3.2.1. Definitions and functional properties

Consider a partition $(\underline{x}_1,\underline{x}_0)$ of the set $\underline{x}$ of variables with $\underline{x}_1=(x_{n-1},\ldots,x_{p+1},x_p)$ and $\underline{x}_0=(x_{p-1},\ldots,x_1,x_0)$; we call *difference of f with respect to* $\underline{x}_0$ the function denoted $\uparrow^{\underline{A}_0} f/\uparrow \underline{x}_0$ and defined as follows :

$$\frac{\uparrow^{\underline{A}_0} f}{\uparrow \underline{x}_0} = \frac{\uparrow^{A_0}}{\uparrow x_0} \left(\frac{\uparrow^{A_1}}{\uparrow x_1} \left(\ldots \frac{\uparrow^{A_{p-1}} f}{\uparrow x_{p-1}}\right) \ldots \right) ; \tag{3.45}$$

$\underline{A}_0$ is the vector $(A_{p-1},\ldots,A_1,A_0)$, $A_i \subseteq \{0,1\}\ \forall\, i$.

In the notation $\uparrow^{\underline{A}_0} f/\uparrow \underline{x}_0$, the difference argument A_i refers to the variable x_i $\forall$i. Starting from the relation (3.1 bis) and using the definition (3.45) iteratively on the variables x_i of $\underline{x}_0$ we obtain the formula :

$$\frac{\uparrow^{\underline{A}_0} f}{\uparrow \underline{x}_0} = \underset{\underline{e}_0}{\uparrow}\ f(x_{p-1} \oplus e_{p-1}x_{p-1}^{(A_{p-1})},\ldots, x_1 \oplus e_1 x_1^{(A_1)}, x_0 \oplus e_0 x_0^{(A_0)}) \,, \tag{3.46}$$

$$\underline{e}_0 = (e_{p-1},\ldots,e_1,e_0), \qquad e_i=0,1 \quad \forall i \qquad .$$

For a two-variable function f(x,y), the formula (3.46) becomes :

$$\frac{\uparrow^{AB} f}{\uparrow xy} = f \uparrow f(x \oplus x^{(A)}) \uparrow f(y \oplus y^{(B)}) \uparrow f(x \oplus x^{(A)},\ y \oplus y^{(B)}) \ . \tag{3.47}$$

The multiple differences enjoy similar properties as the simple differences; these properties were gathered in section 3.1.2.

P.1. The difference $\overset{A_0}{\uparrow} f(\underline{x}_1)/\uparrow\underline{x}_0$ of a function degenerate in $\underline{x}_0$ is a function degenerate in $\underline{x}_0$ (a function degenerate in $\underline{x}_0$ is a function degenerate in each of the variables x_i of $\underline{x}_0$). In particular, the multiple difference of a constant is a constant.

P.2. Adopting the same convention as for the simple differences, we drop in the differential notation the argument A_i when it is {0,1} . The difference $\uparrow f/\uparrow\underline{x}_0$ is degenerate in $\underline{x}_0$; relation (3.46) allows us to write successively :

$$\frac{\uparrow f}{\uparrow \underline{x}_0} = \underset{\underline{e}_0}{\uparrow}\ f(x_{p-1}^{(e_{p-1})},\ldots,x_1^{(e_1)},x_0^{(e_0)})$$

$$= \underset{\underline{e}_0}{\uparrow}\ f(e_{p-1},\ldots,e_1,e_0),\quad e_i=0,1\ \forall i. \tag{3.48}$$

P.3. The *linearity property* (3.9) is satisfied by the difference $\overset{A_0}{\uparrow} f/\uparrow\underline{x}_0$.

P.4. The *duality property* (3.11) is satisfied by the difference $\overset{A_0}{\uparrow} f/\uparrow\underline{x}_0$.

P.5. The operations of difference are commutative, i.e. :

$$\frac{\overset{AB}{\uparrow} f}{\uparrow xy} = \frac{\overset{BA}{\uparrow} f}{\uparrow yx} \ . \tag{3.49}$$

The relation (3.49) derives from the commutativity property of the law $\uparrow$ and from the relation (3.47).

Let us introduce the following notation ; the *ring exponentiation* x^e satisfies the relation :

$$x^e = x \text{ if } e=1 ,$$
$$= 1 \text{ if } e=0 .$$

We shall write :

$$\frac{\overset{A}{\uparrow} f}{\uparrow x^e} = \frac{\overset{A}{\uparrow} f}{\uparrow x} \quad \text{si } e = 1 ,$$
$$= f \quad \text{si } e = 0 ,$$

$$\frac{\overset{\underline{A}_0}{\uparrow} f}{\uparrow \underline{x}_0^{\underline{e}_0}} = \frac{\overset{A_{p-1}\cdots A_1A_0}{\uparrow} f}{\uparrow x_{p-1}^{e_{p-1}}\cdots x_1^{e_1}\, x_0^{e_0}} \ .$$

This convention allows us to write the expression

$$f \uparrow \frac{\overset{A}{\uparrow} f}{\uparrow x} \uparrow \frac{\overset{B}{\uparrow} f}{\uparrow y} \uparrow \frac{\overset{C}{\uparrow} f}{\uparrow z} \uparrow \frac{\overset{AB}{\uparrow} f}{\uparrow xy} \uparrow \frac{\overset{AC}{\uparrow} f}{\uparrow xz} \uparrow \frac{\overset{BC}{\uparrow} f}{\uparrow yz} \uparrow \frac{\overset{ABC}{\uparrow} f}{\uparrow xyz} ,$$

in the form :

$$\underset{e_0e_1e_2}{\uparrow} \quad \frac{\uparrow^{ABC} f}{\uparrow x^{e_0} \; y^{e_1} \; z^{e_2}} , \qquad e_i=0,1 \quad ; \quad i=0,1,2.$$

3.2.2. Expansions and properties of the function $T^{\underline{A}_0} f/T\underline{x}_0$

Using the above notation, we may write the expansion (3.20) in the form :

$$\frac{T^A f}{Tx} = [\underset{e=0,1}{T} \quad ((x^{(A)})^e \textcircled{T} 1) \perp \frac{Tf}{Tx^e}] . \tag{3.50}$$

The definition (3.45) and the formula (3.50) allow us to obtain the expansion of $T^{\underline{A}_0} f/T\underline{x}_0$ in terms of $Tf/T\underline{x}^{\underline{e}_0}$, i.e. :

$$\frac{T^{\underline{A}_0} f}{T\underline{x}_0} = \underset{\underline{e}_0}{T} \; [\underset{i=0,p-1}{\perp} \quad ((x_i^{(A_i)})^{e_i} \textcircled{T} 1) \perp \frac{Tf}{T\underline{x}_0^{\underline{e}_0}}] . \tag{3.51a}$$

With the same symbols as for the simple meet and join differences, the formula (3.51a) gives us the two expansions :

$$\frac{q^{\underline{A}_0} f}{q\underline{x}_0} = \underset{\underline{e}_0}{\vee} [\underset{i=0,p-1}{\wedge} \quad (x_i^{(A_i)})^{e_i} \wedge \frac{qf}{q\underline{x}_0^{\underline{e}_0}}] , \tag{3.51b}$$

$$\frac{p^{\underline{A}_0} f}{p\underline{x}_0} = \underset{\underline{e}_0}{\wedge} [\underset{i=0,p-1}{\vee} \quad \overline{(x_i^{(A_i)})^{e_i}} \vee \frac{pf}{p\underline{x}_0^{\underline{e}_0}}] . \tag{3.51c}$$

The function $x^{(A)}$ being increasing in A (see section 3.1.7) we deduce from (3.51b, 3.51c) that $q^{\underline{A}_0} f/q\underline{x}_0$ is an increasing function in $\underline{A}_0$ while $p^{\underline{A}_0} f/p\underline{x}_0$ is a decreasing function in $\underline{A}_0$. For any pair of vectors $\underline{B}_0=(B_{p-1},\dots,B_1,B_0)$ and $\underline{A}_0=(A_{p-1},\dots,A_1,A_0)$ such that $\underline{B}_0 \geqslant \underline{A}_0$ (i.e. $B_i \geqslant A_i \; \forall i$) we have the following inequalities :

$$\frac{q^{\underline{B}_0} f}{q\underline{x}_0} \geqslant \frac{q^{\underline{A}_0} f}{q\underline{x}_0} \geqslant f \geqslant \frac{p^{\underline{A}_0} f}{p\underline{x}_0} \geqslant \frac{p^{\underline{B}_0} f}{p\underline{x}_0} . \tag{3.52}$$

From the equalities.

$$(A \top b) \top a = (a \top a) \top b = a \top b \quad ,$$

and from the definition (3.46), we deduce that for any pair of vectors :
$\underline{A}_1=(A_{n-1},\ldots,A_{p+1},A_p)$ and $\underline{A}_0=(A_{p-1},\ldots,A_1,A_0)$ we have :

$$\frac{\top^{\underline{A}_1\underline{A}_0} f}{\top \underline{x}_1\underline{x}_0} \top \frac{\top^{\underline{A}_0} f}{\top \underline{x}_0} = \frac{\top^{\underline{A}_1\underline{A}_0} f}{\top \underline{x}_1\underline{x}_0} \quad . \tag{3.53}$$

From the equivalence relation (2.4) of section 2.1.2 and from (3.53), we deduce :

$$\frac{\top^{\underline{A}_1\underline{A}_0} f}{\top \underline{x}_1\underline{x}_0} \perp \frac{\top^{\underline{A}_0} f}{\top \underline{x}_0} = \frac{\top^{\underline{A}_0} f}{\top \underline{x}_0} \quad . \tag{3.54}$$

From the relation :

$$(a \top b) \perp (a \perp b) = a \perp b \quad ,$$

which is generalized in the following way :

$$(\underset{i}{\top}\, a_i) \perp (\underset{i}{\perp}\, a_i) = \underset{i}{\perp}\, a_i \quad ,$$

we deduce :

$$\frac{\top^{\underline{A}_0} f}{\top \underline{x}_0} \perp \frac{\perp^{\underline{A}_0} f}{\perp \underline{x}_0} = \frac{\perp^{\underline{A}_0} f}{\perp \underline{x}_0} \quad . \tag{3.55}$$

An iteration on the number of variables allows us to deduce from (3.26) the expansion of $\top^{\underline{A}_0} f/\top \underline{x}_0$ in terms of the local values of f at $\underline{e}_0$, i.e. :

$$\frac{\top^{\underline{A}_0} f}{\top \underline{x}_0} = \underset{\underline{e}_0}{\top} [\underset{i=0,p-1}{\perp} (x_i^{(e_i \cup A_i)} \circledcirc 1) \perp f(\underline{e}_0)] \, . \tag{3.56}$$

3.2.3. The function δ

We introduce in this section the first of the *composed operators* : these operators are formed from the elementary operators which are the simple and multiple Boolean differences.

Let us first give the expansion of these differences ; from the formula (3.24) we deduce the expansion of $\circledcirc^{\underline{A}_0} f/\circledcirc \underline{x}_0$, i.e. :

$$\frac{\circledcirc^{\underline{A}_0} f}{\circledcirc \underline{x}_0} = \underset{i=0,p-1}{\top} (x_i^{(A_i)} \circledcirc 0) \top \frac{\circledcirc f}{\circledcirc \underline{x}_0} \quad . \tag{3.57}$$

Among the composed operators, the most important is probably the δ-*function*, which we denote $\delta^{\underline{A}_0} f/\delta\underline{x}_0$ and which we define by the relation :

$$\left(\frac{\delta^{\underline{A}_0} f}{\delta\underline{x}_0} ⓣ 1\right) = \underset{\underline{e}_0}{T} \frac{ⓣ^{\underline{A}_0} f}{ⓣ\underline{x}_0^{\underline{e}_0}}, \; e_i=0,1, \; \underline{e}_0 \neq (0,0,\ldots,0) , \tag{3.58}$$

(further on we denote $(0,\ldots,0,0)$ by $\underline{0}$) .
When the operation ⓣ means the equivalence the right-hand side of (3.58) gives us the expression of $\delta^{\underline{A}_0} f/\delta\underline{x}_0$ as a disjunction of differences. The formula (3.25) allows us to verify that the equality (3.58) remains true when ⓣ represents the modulo-2 sum ; the right-hand side of (3.58) is in this case the complement of the function $\delta^{\underline{A}_0} f/\delta\underline{x}_0$.

Relation (3.58) allows us to state that the function δ satisfies the following inequalities :

$$\frac{\delta^{\underline{B}_0} f}{\delta\underline{x}_0} \geqslant \frac{\delta^{\underline{A}_0} f}{\delta\underline{x}_0} \quad \forall\, \underline{B}_0 \geqslant \underline{A}_0 , \qquad \frac{\delta^{\underline{A}_1\underline{A}_0} f}{\delta\underline{x}_1\underline{x}_0} \geqslant \frac{\delta^{\underline{A}_0} f}{\delta\underline{x}_0} . \tag{3.59}$$

Thus the function δ is an increasing function in its differentiation argument. The most important properties of the function δ are state in theorems 3.2.4, 3.2.5 and 3.2.7 below.

An elementary form of the function δ was defined by Akers [1959] :

$$\frac{\delta f}{\delta\underline{x}_0} = \underset{\underline{e}_0}{\vee} \frac{\Delta f}{\Delta\underline{x}_0^{\underline{e}_0}} . \tag{3.60}$$

Akers stated the fundamental property of the function $\delta f/\delta\underline{x}_0$, i.e. : $\delta f/\delta\underline{x}_0$ is degenerate in $\underline{x}_0$; $(\delta f/\delta\underline{x}_0)_{\underline{x}_1=\underline{e}_1}$ is 0 if and only if $f(\underline{e}_1)$ is degenerate in $\underline{x}_0$, otherwise it is 1. The author (Thayse [1972c]) stated the following equalities allowing one to obtain the function $\delta f/\delta\underline{x}_0$ from the prime implicants and from the prime implicates of the function f :

$$\frac{\delta f}{\delta\underline{x}_0} \oplus \frac{pf}{p\underline{x}_0} \oplus \frac{qf}{q\underline{x}_0} = 0 ; \tag{3.61}$$

$$\frac{\delta f}{\delta\underline{x}_0} = \overline{\frac{pf}{p\underline{x}_0}} \; \frac{qf}{q\underline{x}_0} . \tag{3.62}$$

3.2.4. Generalization of theorem 3.1.3.

Theorem

(a) $$\frac{T^{\underline{A}_0} f}{T\underline{x}_0} = \left(\frac{\delta^{\underline{A}_0} f}{\delta \underline{x}_0} \textcircled{T} 1 \right) \textcircled{\perp} \left(\frac{\perp^{\underline{A}_0} f}{\perp \underline{x}_0} \right) \; ; \qquad (3.63)$$

(b) $$\left(\frac{\delta^{\underline{A}_0} f}{\delta \underline{x}_0} \textcircled{T} 1 \right) = \frac{T^{\underline{A}_0} f}{T\underline{x}_0} \perp \overline{\frac{\perp^{\underline{A}_0} f}{\perp \underline{x}_0}} \; . \qquad (3.64)$$

Proof

From relation (3.51) we successively deduce :

$$\frac{T^{\underline{A}_0} f}{T\underline{x}_0} \perp \overline{\frac{\perp^{\underline{A}_0} f}{\perp \underline{x}_0}} = \Big(\mathop{T}_{\underline{e}_0} \Big[\mathop{\perp}_{i} \big((x_i^{(A_i)})^{e_i} \textcircled{T} 1\big) \perp \frac{Tf}{T\underline{x}_0^{\underline{e}_0}} \Big]\Big) \perp$$

$$\Big(\mathop{T}_{\underline{\varepsilon}_0} \Big[\mathop{\perp}_{i} \big((x_i^{(A_i)})^{\varepsilon_i} \textcircled{T} 1\big) \perp \overline{\frac{\perp f}{\perp \underline{x}_0^{\underline{\varepsilon}_0}}} \Big]\Big),$$

$$= \mathop{T}_{\underline{e}_0} \Big[\mathop{\perp}_{i} \big((x_i^{(A_i)})^{e_i} \, T \, 1\big) \perp \Big\{ \Big[\overline{\frac{\perp f}{\perp \underline{x}_0^{\underline{e}_0}}} \perp \Big(\mathop{T}_{\underline{\varepsilon} \subseteq \underline{e}_0} \frac{Tf}{T\underline{x}_0^{\underline{\varepsilon}}} \Big) \Big] \, T$$

$$\Big[\frac{Tf}{T\underline{x}_0^{\underline{e}_0}} \perp \Big(\mathop{T}_{\underline{\varepsilon} \subseteq \underline{e}_0} \overline{\frac{\perp f}{\perp x_0^{\underline{\varepsilon}}}} \Big) \Big] \Big\} \Big],$$

$$= \mathop{T}_{\underline{e}_0} \Big[\mathop{\perp}_{i} \big((x_i^{(A_i)})^{e_i} \textcircled{T} 1\big) \perp \frac{Tf}{T\underline{x}_0^{\underline{e}_0}} \perp \overline{\frac{\perp f}{\perp \underline{x}_0^{\underline{e}_0}}} \Big] \; . \qquad (3.65)$$

The final expression (3.65) derives from simplifications consecutive to the relations :

$$\frac{Tf}{T\underline{x}^{\underline{e}}}\;T\;\frac{Tf}{T\underline{x}^{\underline{\varepsilon}}} = \frac{Tf}{T\underline{x}^{\underline{e}}} \qquad \forall\, \underline{e} \geqslant \underline{\varepsilon}\ .$$

$$\underset{\underline{e}_0}{T}\left(\frac{\text{\textcircled{$\perp$}}^{\underline{A}_0} f}{\text{\textcircled{$\perp$}}\underline{x}_0^{\underline{e}_0}}\right) = \underset{\underline{e}_0}{T}\left[\underset{i}{\perp}\left(\left(x_i^{(A_i)}\right)^{e_i} \text{\textcircled{T}}\ 1\right) \perp \frac{\text{\textcircled{$\perp$}} f}{\text{\textcircled{$\perp$}}\underline{x}_0^{\underline{e}_0}}\right] \quad \text{(from 3.57,3.58))},$$

$$= \underset{\underline{e}_0}{T}\left[\underset{i}{\perp}\left(\left(x_i^{(A_i)}\right)^{e_i} \text{\textcircled{T}}\ 1\right) \perp \frac{Tf}{T\underline{x}_0^{\underline{e}_0}} \perp \overline{\frac{\perp f}{\perp\underline{x}_0^{\underline{e}}}}\right] \quad \text{(from (3.62))}.$$

This proves relation (3.64); relation (3.63) is obtained in a similar way. □

Remark

Formulas (3.63) and (3.64) enjoy an invariance property : they are identical whatever the meaning ($\vee$ or $\wedge$) is given to T. This comes from the fact that we consider the operator δ and not is dual. In the present case the general Boolean notations do not give any new information.

3.2.5. Generalization of theorem 3.1.4.

Theorem

(a) $$f\;T\left(\frac{\delta^{\underline{A}_0} f}{\delta\underline{x}_0} \text{\textcircled{T}}\ 1\right) = \frac{T^{\underline{A}_0} f}{T\underline{x}_0}\ ; \tag{3.66}$$

(b) $$f \perp \left(\frac{\delta^{\underline{A}_0} f}{\delta\underline{x}_0} \text{\textcircled{T}}\ 1\right) = f\ \text{\textcircled{$\perp$}}\ \frac{\perp^{\underline{A}_0} f}{\perp\underline{x}_0}\ ; \tag{3.67}$$

(c) $$f = \underset{\underline{e}_0}{\perp}\ \frac{T^{\underline{A}_0 (\underline{e}_0)} f}{T\underline{x}_0}\ , \qquad e_i \in \{0,1\}\ . \tag{3.68}$$

Proof

We successively deduce from relation (3.64) :

(a) $$f\;T\left(\frac{\delta^{\underline{A}_0} f}{\delta\underline{x}_0} \text{\textcircled{T}}\ 1\right) = f\;T\left(\frac{T^{\underline{A}_0} f}{T\underline{x}_0} \perp \overline{\frac{\perp^{\underline{A}_0} f}{\perp\underline{x}_0}}\right),$$

$$= \left(f\;T\,\frac{T^{\underline{A}_0} f}{T\underline{x}_0}\right) \perp \left(f\;T\,\overline{\frac{\perp^{\underline{A}_0} f}{\perp\underline{x}_0}}\right),$$

$$= \frac{T^{\underline{A}_0} f}{T\underline{x}_0}\ .$$

This last equality derives from the relations

$$f \top \frac{\top^{\underline{A}_0} f}{\top \underline{x}_0} = \frac{\top^{\underline{A}_0} f}{\top \underline{x}_0} \text{ and } f \top \overline{\frac{\perp^{\underline{A}_0} f}{\perp \underline{x}_0}} \tag{3.69}$$

which themselves are obtained from the equality (3.51a) .

(b)

$$f \perp \left(\frac{\delta^{\underline{A}_0} f}{\delta \underline{x}_0} \mathbin{\textcircled{$\top$}} 1\right) = f \perp \frac{\top^{\underline{A}_0} f}{\top \underline{x}_0} \perp \overline{\frac{\top^{\underline{A}_0} f}{\top \underline{x}_0}} ,$$

$$= f \perp \overline{\frac{\perp^{\underline{A}_0} f}{\perp \underline{x}_0}} . \tag{3.70a}$$

This last equality derives from the relation

$$f \perp \frac{\top^{\underline{A}_0} f}{\top \underline{x}_0} = f ,$$

which is itself obtained from (3.54). The Stone theorem (relation 2.13) and the absorption laws allow us to write successively :

$$f \mathbin{\textcircled{$\perp$}} \frac{\perp^{\underline{A}_0} f}{\perp \underline{x}_0} = \left(f \perp \overline{\frac{\perp^{\underline{A}_0} f}{\perp \underline{x}_0}}\right) \top \left(\bar{f} \perp \frac{\perp^{\underline{A}_0} f}{\perp \underline{x}_0}\right) ,$$

$$= \left(f \perp \overline{\frac{\perp^{\underline{A}_0} f}{\perp \underline{x}_0}}\right) \top e_{\top} ,$$

$$= f \perp \overline{\frac{\perp^{\underline{A}_0} f}{\perp \underline{x}_0}} . \tag{3.70b}$$

The proof of part (b) derives from a comparison between the relations (3.70a) and (3.70b).

(c) The Boolean exponentiation for a set being defined in the same way as for an element, i.e. :

$$A_i^{(e_i)} = A_i \text{ if } e_i = 1 ,$$

$$= \bar{A}_i \text{ if } e_i = 0 ,$$

and the notation $\underline{A}_0^{(\underline{e}_0)}$ meaning $(A_{p-1}^{(e_{p-1})}, \ldots, A_1^{(e_1)}, A_0^{(e_0)})$, the relation (3.68) is deduced from (3.16) by perfect induction on the number of variables.

3.2.6. Functional properties of the meet and join differences

We adopt for the multiple differences the same callings and notations as for the simple differences ; $p^{\underline{A}_0} f/p\underline{x}_0$ and $q^{\underline{A}_0} f/q\underline{x}_0$ are the meet and join differences with respect to $\underline{x}_0$ respectively. When the vector $\underline{A}_0$ is formed only by scalars we shall denote it (as usual) by $\{\underline{h}_0\} = \{h_{p-1},\dots,h_1,h_0\}$, $h_i \in \{0,1\}$ $\forall i$. The differences $p^{\{\underline{h}_0\}} f/p\underline{x}_0$ and $q^{\{\underline{h}_0\}} f/q\underline{x}_0$ are also called *lower* and *upper envelopes* respectively. The most important functional properties of the meet and join differences are gathered in the statement of the theorem below.
We say that a function is *$\underline{A}_0$-degenerated in $\underline{x}_0$* if it is A_i-degenerated in x_i, $\forall i$, $\underline{A}_0=(A_{p-1},\dots, A_1,A_0)$, $\underline{x}_0=(x_{p-1},\dots,x_1,x_0)$.

Theorem

The function $p^{\underline{A}_0} f/p\underline{x}_0$ *(resp.* $q^{\underline{A}_0} f/q\underline{x}_0$*) is a function $\underline{A}_0$-degenerate in $\underline{x}_0$ equal to the disjunction of the prime implicants of f (resp. of $\bar{f}$) $\underline{A}_0$-degenerate in $\underline{x}_0$; it is the greatest function $\underline{A}_0$-degenerate in $\underline{x}_0$ and smaller than or equal to* f *(resp.* $\bar{f}$*).*

Proof

The function $p^{\underline{A}_0} f/p\underline{x}_0$ is $\underline{A}_0$-degenerate in $\underline{x}_0$ (see 3.51c)) and is smaller than or equal to f (see 3.46)) ; let us show that this function is the disjunction of the prime implicants of f $\underline{A}_0$-degenerate in $\underline{x}_0$. Any prime implicant of f $\underline{A}_0$-degenerate in $\underline{x}_0$ is of the form :

$$x_j^{(\bar{A}_j)}\, x_k^{(\bar{A}_k)} \dots x_\ell^{(\bar{A}_\ell)}\; g(\underline{x}_1),\ (i,j,\dots,k)\subseteq(p-1,\dots,1,0)\ ,$$

where $g(\underline{x}_1)$ is a cube containing only the letters $x_i \in \underline{x}_1$.
Moreover, the expression :

$$x_j^{(\bar{A}_j)}\, x_k^{(\bar{A}_k)} \dots x_\ell^{(\bar{A}_\ell)}\; g(\underline{x}_1)\; \overline{\frac{p^{\underline{A}_0} f}{p\underline{x}_0}} =$$

$$x_j^{(\bar{A}_j)}\, x_k^{(\bar{A}_k)} \dots x_\ell^{(\bar{A}_\ell)}\; g(\underline{x}_1)\; \{\bigvee_{\underline{e}_0} [\bigwedge_i (x_i^{(A_i)})^{e_i}\; \overline{\frac{p f}{p\underline{x}_0^{\underline{e}_0}}}\,]\}\ ,$$

is identically zero. Indeed, since $x_i^{(\bar{A}_i)}\, x_i^{(A_i)} \equiv 0$, the terms which are not evidently zero are of the form :

$$x_j^{(\bar{A}_j)}\, x_k^{(\bar{A}_k)} \dots x_\ell^{(\bar{A}_\ell)}\; g(\underline{x}_1)\; \{\bigvee_{\underline{e}_0^*} [\bigwedge_i (x_i^{(A_i)})^{e_i}\; \overline{\frac{p f}{p\underline{x}_0^{\underline{e}_0^*}}}\,]\ \}$$

with $\underline{e}_0^* = \underline{e}_0 \setminus (e_j, e_k, \ldots, e_\ell)$. This last expression is identically zero since

$$\frac{pf}{p\underline{x}_0^{\underline{e}_0^*}} \geq x_j^{(\bar{A}_j)} x_k^{(\bar{A}_k)} \ldots x_\ell^{(\bar{A}_\ell)} g(\underline{x}_1) .$$

Indeed, $pf/p\underline{x}_0^{\underline{e}_0^*}$ is the disjunction of all the prime implicants of f being degenerate in $\underline{x}_0^{\underline{e}_0^*}$ (Thayse [1973a] ,[1976]) and $x_j^{(\bar{A}_j)} \ldots x_\ell^{(\bar{A}_\ell)} g(\underline{x}_1)$ is an implicant degenerate in $\underline{x}_0^{\underline{e}_0^*}$. The function $p^{\underline{A}_0} f/p\underline{x}_0$ being greater than or equal to any prime implicant of f $\underline{A}_0$-degenerate in $\underline{x}_0$, it is also greater than or equal to their disjunction. On the other hand, by definition of the concept of prime implicant, the disjunction of all the prime implicants of f $\underline{A}_0$-degenerate in $\underline{x}_0$ is the greatest function smaller than or equal to f and $\underline{A}_0$-degenerate in $\underline{x}_0$. Indeed, if there was a greater function, at least one of its prime implicants would be greater and it would also be an implicant of f, which is impossible. This proves the theorem for the functions f and $p^{\underline{A}_0} f/p\underline{x}_0$; a dual type of proof holds for the functions $\bar{f}$ and $\overline{q^{\underline{A}_0} f/q\underline{x}_0}$. □

3.2.7. <u>Properties of the function δ</u>

From theorem 3.2.6 and from relation (3.64) we deduce the following property which characterizes the function δ .

<u>Theorem</u>

The function $\delta^{\underline{A}_0} f/\delta\underline{x}_0$ *is a function* $\underline{A}_0$*-degenerate in* $\underline{x}_0$ *, equal to the disjunction of the prime implicants of* f *and of* $\bar{f}$ $\underline{A}_0$*-degenerate in* $\underline{x}_0$.

In particular if f is degenerate in $\underline{x}_0$, so is also $\bar{f}$ and hence all the prime implicants of f and $\bar{f}$ are degenerate in $\underline{x}_0$. It follows that $\overline{\delta f/\delta\underline{x}_0}$ is in this case the disjunction of f and of $\bar{f}$ i.e. 1. conversely, if $\overline{\delta f/\delta\underline{x}_0} = 1$ the disjunction of the prime implicants (degenerate in $\underline{x}_0$) of f and of $\bar{f}$ covers the complete domain function. All the prime implicants of f and of $\bar{f}$ are thus degenerate in $\underline{x}_0$ and so are also f and $\bar{f}$.

Thus we obtain as a particular case of theorem 3.2.7 a proposition due to Akers [1959] quoting that : a function f is degenerate in $\underline{x}_0$ if and only if $\delta f/\delta\underline{x}_0$ is zero.

In the same way as $\delta f/\delta\underline{x}_0$ is connected to the property of degeneracy , the function $\delta^{\underline{A}_0} f/\delta\underline{x}_0$ is connected to the property of $\underline{A}_0$-degeneracy. We show in section 3.2.16 the following proposition :

The function f is $\underline{A}_0$-degenerate in $\underline{x}_0$ if and only if :

$$f \frac{\delta^{\underline{A}_0} f}{\delta \underline{x}_0} \equiv 0 ,$$

where "≡" means an identity which holds for each value of $\underline{x}_0$.

3.2.8. Applications in switching theory

We give here a short review of the use that may be done in switching theory of some of the concepts and theorems quoted above. This review will be short and thus incomplete : instead of developing computation algorithms in order to solving some problems in an optimal way our purpose will only be to interpret the developed formalism in order to give rise to a further more detailed and more applied study. Theorems 3.2.4 and 3.2.5 state properties connecting the meet and join differences and the function δ. Theorems 3.2.6 and 3.2.7 develop some functional properties of the meet and join differences and of the function δ. The function δ constitutes the ground function for several algorithms allowing us to detect the *decomposability* of a function. The fact that a function undergoes some type of decomposition is certainly an important functional property as illustrated by the following classical example : we say that a function $f(\underline{x})$ is decomposable according to the partition $(\underline{x}_1, \underline{x}_0)$ of the variables if it may be written in the form :

$$f(\underline{x}) = F[\underline{x}_1, y(\underline{x}_0)] \tag{3.71}$$

where F and y are also switching functions. In this case, f is entirely defined by the truth table of F that contains 2^{n-p+1} rows, by the truth table of y that contains 2^p rows and by the knowledge that f undergoes the given decomposition ; thanks to that knowledge, the number of bits to memorize in order to define f completely is $(2^{n-p+1}+2^p)$ instead of 2^n. Now, if one realizes f by means of read only memory (R.O.M.) circuits, the saving in the number of bits necessary for defining f corresponds to an actual hardware saving in the realization of f. The following proposition has been stated by the author(Thayse [1972c]) :
A function f is decomposable according to the form (3.71) if and only if it satisfies the relations :

$$\frac{\Delta f}{\Delta x_i} = \frac{\delta f}{\delta \underline{x}_0} g_i(\underline{x}_0) , \quad \forall x_i \in \underline{x}_0 . \tag{3.72}$$

The function δ is also used in several algorithms related to *hazard detection*. We say that a transition between two vertices which differ from p bits (the concept of distance on the n-cube is the Hamming distance), is *hazard-free* if the function f changes monotonically when the p+1 vertices of any path connecting these two vertices are runned over. The literature classifies the function hazards into *static hazards* and *dynamic hazards*. The characteristic of the static hazard is that it causes a transition in a function which is assumed to remain constant, during a given input variable change. The dynamic hazard, which can occur when the function is meant to change, causes the function to change three or more times instead of only once. The following proposition has been stated by the author (Thayse [1971]) :

A transition between the vertices $(\underline{x}_1, \underline{x}_0) = (\underline{a}_1, \underline{a}_0)$ *and* $(\underline{x}_1, \underline{x}_0) = (\underline{a}_1, \bar{\underline{a}}_0)$ *gives rise to a static hazard for the function* f *if and only if the following condition holds :*

$$\left(f \oplus f(\bar{\underline{x}}_0) \oplus \frac{\delta f}{\delta \underline{x}_0}\right)_{\underline{x}_1=\underline{a}_1, \underline{x}_0=\underline{a}_0} = 1 \, . \tag{3.73}$$

We have shown two applications of the function δ, the decomposability detection and the hazard detection; other applications of this function will be analyzed further on.

The concepts of implicants and implicates are classical and their utility has no longer to be demonstrated : several functional properties such as the decomposability (Deschamps [1975]), the unateness (Harrison [1965], pp. 187-191) and the symmetry (Deschamps [1973]) may be made obvious by the analysis of corresponding properties of the prime implicants or of the prime implicates. These last concepts are moreover the foundation for some types of optimal two-level synthesis. Relations (3.64) and (3.65) allow us to compute the δ functions from the knowledge of the prime implicants and of the prime implicates and conversely.

At this point it is useful to present a first connection of the meet and join differences and hence of the δ function with the consensus theory. A well-known procedure for finding prime implicants of a switching function is based on what Quine [1955] first called the *consensus* of implicants. The original method developed by Quine is generally referred to as the *iterative consensus;* it is fully detailed in most of the books dealing with switching theory (see e.g. Mc Cluskey [1965]). This method has been improved, by Tison [1965, 1967, 1971] who suggested a more efficient algorithm which is called the *generalized consensus*. The author has shown (Thayse [1978]) that the meet and join differences and hence the δ function are a convenient mathematical tool for both the iterative and the generalized consensus. In this way we shall show in chapter 4 that the consensus algorithms are implicitly contained in the meet and join difference algorithms developed by the author (Thayse [1978]).

The functions δ may also be evaluated by starting e.g. from the discrete Fourier transform (Davio, Thayse and Bioul [1972]) or from the truth table of f. Detailed algorithms concerning this evaluation may be found in Thayse [1972c], Thayse and Davio [1973], Thayse [1975a], Davio, Deschamps and Thayse [1978] .

Let us finally give an interpretation in terms of switching theory of relation (3.68); when $\underline{x}_0$ is the whole set $\underline{x}$ of variables, we write this relation in the form :

$$f = \underset{\underline{e}}{\perp} \frac{T^{\underline{A}^{(\underline{e})}} f}{T\underline{x}} \;,\; \underline{e} = (e_{n-1},\dots,e_1,e_0),\; e_i \in \{0,1\};$$

$$\underline{A} = (A_{n-1},\dots,A_1,A_0),\; A_i \subseteq \{0,1\}. \tag{3.74}$$

If we give several meanings to both the operations T and the arguments A_i, the expression (3.74) allows us to state the various entries of Table 3.II.

T \ A_i	{0,1} or ∅	{0} or {1}
∨	The function f is the disjunction of all its implicants	The function f is the disjunc- of all its lower envelopes
∧	The function f is the conjunc- of all its implicates	The function f is the conjunction of all its upper envelopes

Table 3.II.

When the A_i are all 0,1 or ∅ (these two values are equivalent since the formula (3.74) is symmetric in $A_i, \bar{A}_i$) and if we take for T the conjunction we obtain the disjunction of all the implicants of f from which we deduce the prime implicants by absorption. When the A_i are all 0 or 1 we obtain the function f as a disjunction of all its *lower envelopes* ; in agreement with the definitions 3.1.9, the differences $p^{\{\underline{h}\}}f/p\underline{x}$ are also called lower envelopes. We may obtain the *prime lower envelopes* from (3.74) by application of the absorption law. In the same way, as the representation of f by the disjunction of its prime implicants is in one-to-one correspondence with two-level networks, the representation of f as disjunction of its first lower envelopes in one-to-one correspondence with three-level networks (Thayse and Deschamps [1977], see also chapter 4). When we give to T the dual interpretation of disjunction we may associate to (3.74) dual statements. We obtain f, either as the conjunction of its (prime) implicates or as the conjunction of its (*prime*) *upper envelopes*, these last being the differences $q^{\{\underline{h}\}}f/q\underline{x}$.

3.2.9. Boolean differences and the Galoisian expansions.

In the preceding sections we studied some functional properties of the functions $T^{\underline{A}_0}f/T\underline{x}_0$ and of the function δ . We consider in this section and in the following ones the functions $\oplus f/\oplus\underline{x}_0$.

From relation (3.57) we deduce :

$$\frac{\textcircled{T}^{A_0} f}{\textcircled{T} \underline{x}_0} = \overline{\frac{\textcircled{I}^{A_0} f}{\textcircled{I} \underline{x}_0}} .$$

Further on we study only the differences associated with the modulo-2 sum, those associated with the equivalence being deduced by simple complementation. Akers [1959] showed that any Boolean function has the expansion :

$$f = \bigoplus_{\underline{e}_0} \left(\frac{\Delta f}{\Delta \underline{x}_0^{\underline{e}_0}}\right)_{\underline{x}_0=\underline{0}} \left[\bigwedge_{i=0,p-1} x_i^{e_i}\right] , \; e_i \in \{0,1\} \; \forall i \tag{3.75}$$

Davio and Piret [1969] proved the following generalization of (3.75).

Theorem (Newton expansion with respect to the variables $\underline{x}_0$)

Any Boolean function has the expansion

$$f = \bigoplus_{\underline{e}_0} \left(\frac{\Delta f}{\Delta \underline{x}_0^{\underline{e}_0}}\right)_{\underline{x}_0=\underline{h}_0} \left[\bigwedge_{i=0,p-1} (x_i \oplus h_i)^{e_i}\right]$$

$$\underline{h}_0 = (h_{p-1},\ldots,h_1,h_0), \; h_i \in \{0,1\} \; \forall \, i . \tag{3.76}$$

We call (3.76) the Newton expansion of f at $\underline{x}_0=\underline{h}_0$; expansion (3.75) is then nothing but the Newton expansion of f at $\underline{x}_0=\underline{0}$. The expansions (3.75) and (3.76) may also be obtained from (2.47) or (2.48) : we have to interpret the extended state vector $\underline{\phi}^{(\textcircled{T})}(f)$ in terms of the local values of the differences $\textcircled{T}f/\textcircled{T}\underline{x}^{\underline{e}}$, i.e. :

$$f = \textcircled{T}_{\underline{e}} \left\{\left(\frac{\textcircled{T}f}{\textcircled{T}\underline{x}^{\underline{e}}}\right)_{\underline{x}=\underline{h}} \; \mathrm{T}\left[\mathop{\mathrm{T}}_{i=0,n-1} ((x_i \oplus h_i)^{e_i} \textcircled{I} \, 1)\right]\right\} . \tag{3.77}$$

The expansions (3.76) and (3.77) coincide when the operation $\textcircled{T}$ is the modulo-2 sum and when $\underline{x}_0$ is the complete set of variables ; further on the calling *Newton expansion* will mean the expansion (3.76) according to $\underline{x}$. Observe that (3.77) may be obtained from (3.38) by means of an induction on the number of variables. Let us finally quote the most important functional property of the Boolean difference.

In view of relation (3.48) it is clear that the function $\Delta f/\Delta \underline{x}_0$ is degenerate in $\underline{x}_0$; more precisely $(\Delta f/\Delta \underline{x}_0)_{\underline{x}_1=\underline{e}_1}$ is 0 only if the function $f(\underline{e}_1)$ has an even number of 1 in the subcube $\underline{x}_1=e_1$, otherwise $(\Delta f/ \Delta \underline{x}_0)_{\underline{x}_1=\underline{e}_1}$ is 1.

We deduce that the function

$$\frac{\Delta^{\underline{A}_0} f}{\Delta \underline{x}_0} = \bigwedge_i x_i^{(A_i)} \frac{\Delta f}{\Delta \underline{x}_0}$$

is $\bar{\underline{A}}_0$-degenerate in $\underline{x}_0$; $(\Delta^{\underline{A}_0} f/\Delta \underline{x}_0)_{\underline{x}_1=\underline{e}_1}$ is zero if $f(\underline{e}_1)$ has an even number of 1, otherwise it is equal to $\wedge x_i^{(A_i)}$.

3.2.10. The Boolean differences and the circuit optimization.

We observed that representations of f as disjunction of prime implicants or as disjunction of prime lower envelopes are respectively structure formulas for 2-level and 3-level networks using the gates OR and AND. In the same way the Newton expansions (3.76) are structure formulas for 2-level realizations of networks using AND and EXCLUSIVE-OR gates. The network represented by the formula (3.76) is formed by 1 EXCLUSIVE-OR gate and as many AND-gates as non-zero terms $(\Delta f/\Delta \underline{x}^{\underline{e}})_{\underline{x}=\underline{h}}$. The interest, in the optimization of switching circuits, of minimizing the number of non-zero terms $(\Delta f/\Delta \underline{x}^{\underline{e}})_{\underline{x}=\underline{h}}$ then immediately appears : we minimize at the same time the number of AND gates necessary to build the function f. This minimization is obtained by choosing the vertex (vertices) $\underline{h}$ where a maximum number of terms $(\Delta f/\Delta \underline{x}^{\underline{e}})_{\underline{x}=\underline{h}}$ are zero ; stated otherwise, we have to minimize the function $\sum_{\underline{e}} \Delta f/\Delta \underline{x}^{\underline{e}}$, where $\sum$ means the sum.

On a vertex $\underline{h}$, each of the terms $\Delta f/\Delta \underline{x}^{\underline{e}}$ has either the value 0 or the value 1; the minimization of the sum of these terms may thus be obtained by using a pseudo-Boolean method (Hammer and Rudeanu [1968], chapter VI, Thayse [1974a]).

3.2.11. The sensitivities

We introduced in section 3.2.3 a first composed operator and we called it *δ function*. We introduce here the composed operator called *sensitivity* . We denote by E^A/Ex the operator *displacement* with respect to x and we define it as follows :

$$\frac{E^A f}{Ex} = f(x \oplus x^{(A)}) \ . \tag{3.78}$$

We use the following symbolic notation to define the difference of f with respect to x :

$$\frac{\Delta^A f}{\Delta x} = f(x \oplus x^{(A)}) \oplus f = \frac{E^A f}{Ex} \oplus f = (\frac{E^A}{Ex} \oplus 1) f \tag{3.79}$$

The formula (3.45) allows us to write $\Delta^{\underline{A}_0} f/\Delta \underline{x}_0$ in the following way :

$$\frac{\Delta^{\underline{A}_0} f}{\Delta \underline{x}_0} = \bigwedge_{i=0,p-1} \left(\frac{E^{A_i} f}{E x_i} \oplus 1 \right) f \ . \tag{3.80}$$

Be $[\Delta^{\underline{A}_0} f / \Delta \underline{x}_0^{\underline{e}_0}]$ the $(2^p \times 1)$ matrix of differences obtained by giving to each of the $e_i \in \underline{e}_0$ the values 0 and 1 ; the differences are gathered in this matrix in lexicographic order, i.e. $\Delta^{\underline{A}_0} f / \Delta \underline{x}^{\underline{e}_0}$ is at the $(2^0 e_0 + 2^1 e_1 + \ldots + 2^{p-1} e_{p-1})$-th row. The $(2^p \times 1)$ matrix of functions $E^{\underline{A}_0} f / E \underline{x}_0^{\underline{e}_0}$ will be denoted $[E^{\underline{A}_0} f / E \underline{x}_0^{\underline{e}_0}]$. If A is the $(2^p \times 2^p)$ matrix formed by the binomial coefficients evaluated modulo 2 we have (Calingaert [1961]) :

$$[\Delta^{\underline{A}_0} f / \Delta \underline{x}_0^{\underline{e}_0}] = A[\oplus \wedge] \, [E^{\underline{A}_0} f / E \underline{x}_0^{\underline{e}_0}] \ . \tag{3.81}$$

The matrix A being its own inverse, we have also

$$\begin{aligned} &[E^{\underline{A}_0} f / E \underline{x}_0^{\underline{e}_0}] = A \, [\oplus \wedge] \, [\Delta^{\underline{A}_0} f / \Delta \underline{x}^{\underline{e}_0}] \ , \\ &\frac{E^{\underline{A}_0} f}{E \underline{x}_0} = f(\underline{x}_0 \oplus \underline{x}^{(\underline{A}_0)}) = \bigoplus_{\underline{e}_0} \frac{\Delta^{\underline{A}_0} f}{\Delta \underline{x}_0^{\underline{e}_0}} \ , \ e_i = 0,1 \ . \end{aligned} \tag{3.82}$$

The *sensitivity* of f with respect to $\underline{x}_0$, which we denote $S^{\underline{A}_0} f / S \underline{x}_0$, is the function defined by the following relation :

$$\frac{S^{\underline{A}_0} f}{S \underline{x}_0} = f \oplus f(\underline{x}_0 \oplus \underline{x}_0^{(A_0)}) \ . \tag{3.83}$$

From (3.81) and (3.82) we deduce that the Boolean differences and the sensitivities are related by the expressions (3.84,3.85).

<u>Theorem</u>

$$\frac{\Delta^{\underline{A}_0} f}{\Delta \underline{x}_0} = \bigoplus_{\underline{e}_0} \frac{S^{\underline{A}_0} f}{S \underline{x}_0^{e_0}} \ , \ e_i = 0,1; \ \underline{e}_0 \neq 0 \ ; \tag{3.84}$$

$$\frac{S^{\underline{A}_0} f}{S \underline{x}_0} = \bigoplus_{\underline{e}_0} \frac{\Delta^{\underline{A}_0} f}{\Delta \underline{x}_0^{\underline{e}_0}} , \ e_i = 0,1; \ \underline{e}_0 \neq 0 \ . \tag{3.85}$$

In view of the definition of the sensitivity, it is obvious that this function does not change when $\underline{x}_0 \oplus \underline{x}_0^{(\underline{A}_0)}$ is substituted for $\underline{x}_0$. This property will be used further on.

3.2.12. Theorem

Any Boolean function f has the expansions :

(a) $$f=f(\underline{x}_0^{(\underline{A}_0)}) \oplus \sum_{\underline{e}_0} \frac{\Delta f}{\Delta \underline{x}_0^{\underline{e}_0}} [\bigwedge_{i=0,p-1} (x_i^{(A_i)} \oplus x_i)^{e_i}] ; \qquad (3.86)$$

(b) $$f=f(\underline{x}_0^{(A_0)}) \oplus \sum_{\underline{e}_0} (\frac{\Delta f}{\Delta \underline{x}_0^{\underline{e}_0}})_{\underline{x}_0=\underline{x}_0^{(A_0)}} [\bigwedge_{i=0,p-1} (x_i^{(A_i)} \oplus x_i)^{e_i}] . \qquad (3.87)$$

From (3.85) we deduce :

$$f = f(\underline{x}_0 \oplus \underline{x}_0^{(A_0)}) \oplus \sum_{\underline{e}_0} \frac{\Delta^{(A_0)} f}{\Delta \underline{x}_0^{\underline{e}_0}}$$

$$= f(\underline{x}_0 \oplus \underline{x}_0^{(A_0)}) \oplus \sum_{\underline{e}_0} \frac{\Delta f}{\Delta \underline{x}_0^{\underline{e}_0}} [\bigwedge_{i=0,p-1} (x_i^{(A_i)})^{e_i}] .$$

If in this last expression we substitute $x_i^{(A_i)} \oplus x_i$ for x_i $\forall x_i \in \underline{x}_0$, we obtain the relation :

$$f(\underline{x}_0^{(A_0)}) = f \oplus \sum_{\underline{e}_0} \frac{\Delta f}{\Delta \underline{x}_0^{\underline{e}_0}} [\bigwedge_{i=0,p-1} (x_i^{(A_i)} \oplus x_i)^{e_i}] , \qquad (3.38)$$

which is another form of (3.86).

If $x_i^{(A_i)}$ and x_i are reversed $\forall x_i \in \underline{x}_0$, we obtain the expansion (3.87). □

The expansion (3.87) is nothing but the Newton expansion; this expansion has been written together with (3.86) in order to make obvious the fact that it is regardless in these expressions to evaluate the differences at the vertices $\underline{x}_0$ or $\underline{x}^{(A_0)}$. The formulas (3.86) and (3.87) provide us also with corresponding expansions for the sensitivities ; relation (3.86) allow us to write :

$$f \oplus f(\underline{x}_0^{(\underline{A}_0)}) = \frac{S^{\underline{A}_0 \nabla \{\underline{1}\}} f}{S\underline{x}_0} \quad , \tag{3.89}$$

where according to the usual vectorial notation, $\underline{A}_0 \nabla \{\underline{1}\}$ means $(A_{p-1} \nabla 1,\ldots,A_1 \nabla 1, A_0 \nabla 1)$.

3.2.13. Meaning of the sensitivity function

A function f is *invariant with respect to* $\underline{x}_0$ if it does not change when $\underline{\bar{x}}_0$ is substituted for $\underline{x}_0$, i.e. if is satisfies the relation $f=f(\bar{\underline{x}}_0)$, or, stated otherwise if :

$$\frac{Sf}{S\underline{x}_0} = 0 \ . \tag{3.90}$$

We shall give an interpretation of the sensitivity function in terms of fault detection (see also section 3.1.8.).

A *fault* or *failure* of an electrical circuit is a physical defect of one or several components, which can cause the circuit to malfunction. Many faults in electrical circuits create a *stuck-at-fault* in the corresponding switching network : in this kind of fault model, it is assumed that any electrical fault can be modelled by a number of connections in the corresponding logical network permanently fixed at the logical level 0 or 1. The main purpose of fault detection is to *detect* faults in logic networks.

Consider a logic network realizing a function $f(\underline{x})$; assume further that the network, undergoing a specific fault, realizes the function $g(\underline{x})$ instead of $f(\underline{x})$.

Any n-tuple $\underline{h}$ of particular values of the input variables $\underline{x}$ is called *test-vector* for this specific fault if and only if $f(\underline{h}) \neq g(\underline{h})$.
Consider the function :

$$t(\underline{x}) = f(\underline{x}) \oplus g(\underline{x}) \ . \tag{3.91}$$

This function assumes a value different from zero if a test-vector h is substituted for $\underline{x}$. The function $t(\underline{x})$ will be called hereafter *test-function* for that specific fault.

Theorem

(a) *The sensitivity* $S^{\underline{A}_0} f/S\underline{x}_0$ *is the test function for the fault :* $\underline{x}_0$ *is stuck at* $x_0 \oplus \underline{x}_0^{(\underline{A}_0)}$, $A_i \subseteq \{0,1\}$ $\forall$ i.

(b) *The function* $\delta^{\underline{A}_0} f/\delta\underline{x}_0$ *is the test function for at least one of the faults : a subset of* $\underline{x}_0$ *is stuck-at* $x_i \oplus x_i^{(A_i)}$ $\forall x_i$ *belonging to that subset.*

Proof

(a) If $\underline{x}_0$ is stuck-at $\underline{x}_0^{(\underline{A}_0)}$, the function g of the definition (3.91) is $f(\underline{x}_0 \oplus \underline{x}_0^{(\underline{A}_0)})$ and the test-function is thus

$$t = f \oplus f(\underline{x}_0 \oplus \underline{x}_0^{(\underline{A}_0)}) = \frac{S^{\underline{A}_0} f}{S\underline{x}_0} .$$

(b) Part (b) derives from part (a) and from the following formula that will be proved below :

$$\frac{\delta^{\underline{A}_0} f}{\delta \underline{x}_0} = \bigvee_{\underline{e}_0} \frac{S^{\underline{A}_0} f}{S\underline{x}_0^{\underline{e}_0}} , \quad \underline{e}_0 \neq \underline{0} \tag{3.92}$$

From the definition of sensitivity we successively deduce :

$$\frac{S^{\underline{A}_0} f}{S\underline{x}_0} = f \oplus f(\underline{x}_0 \oplus \underline{x}_0^{(\underline{A}_0)}) ,$$

$$\bigvee_{\underline{\varepsilon}_0} [f \oplus f(\underline{x}_0^{(\bar{\underline{\varepsilon}}_0)})] [\bigwedge_i (x_i^{(A_i)})^{(\varepsilon_i)}] ,$$

$$= \bigvee_{\underline{\varepsilon}_0} \frac{Sf}{S\underline{x}_0^{\underline{\varepsilon}_0}} [\bigwedge_i (x_i^{(A_i)})^{(\varepsilon_i)}] , \ \varepsilon_i \in \{0,1\} , \ \underline{\varepsilon}_0 \neq 0 .$$

$$\bigvee_{\underline{e}_0} \frac{S^{\underline{A}_0} f}{S\underline{x}_0^{\underline{e}_0}} = \bigvee_{\underline{e}_0, \underline{\varepsilon}_0} \frac{Sf}{S\underline{x}_0^{\underline{\varepsilon}_0}} [\bigwedge_i (x_i^{(A_i)})^{(\varepsilon_i)}] , \qquad \underline{\varepsilon}_0 \subseteq \underline{e}_0 ,$$

$$= \bigvee_{\underline{\varepsilon}_0} [\bigvee_{\underline{e}_0} \frac{Sf}{S\underline{x}_0^{\underline{e}_0}}] [\bigwedge_i (x_i^{(A_i)})^{(\varepsilon_i)}] , \qquad \underline{e}_0 \subseteq \underline{\varepsilon}_0 .$$

Taking into account that (Davio et Piret [1969]) :

$$\frac{\delta f}{\delta \underline{x}_0} = \bigvee_{\underline{e}_0} \frac{Sf}{S\underline{x}_0^{\underline{e}_0}} = \bigvee_{\underline{e}_0} \frac{\Delta f}{\Delta \underline{x}_0^{\underline{e}_0}} , \tag{3.93}$$

we may write the last equality :

$$\bigvee_{\underline{e}_0} \frac{S^{\underline{A}_0} f}{S\underline{x}_0^{\underline{e}_0}} = \bigvee_{\underline{\varepsilon}_0} \frac{\delta f}{\delta \underline{x}_0^{\underline{\varepsilon}_0}} \left[\bigwedge_i \left(x_i^{(A_i)}\right)^{(\varepsilon_i)} \right] ,$$

$$= \bigvee_{\underline{\varepsilon}_0} \frac{\Delta f}{\Delta \underline{x}_0^{\underline{\varepsilon}_0}} \left[\bigwedge_i \left(x_i^{(A_i)}\right)^{\varepsilon_i} \right] = \frac{\delta^{\underline{A}_0} f}{\delta \underline{x}_0} \quad . \qquad \square$$

3.2.14 Meaning of the function $Sf/S\underline{x}_0 \oplus \delta f/\delta\underline{x}_0$

In sections 3.2.8 and 3.2.12 we interpreted several functions such as the Boolean difference, the sensitivities, the δ functions and the meet and join differences in terms of problems arising in switching theory. We continue this interpretation of the Boolean operators by analyzing the function $(Sf/S\underline{x}_0 \oplus \delta f/\delta\underline{x}_0)$. First of all we observe that the condition (3.72) may also be written in the form :

$$\left(\frac{Sf}{S\underline{x}_0} \oplus \frac{\delta f}{\delta \underline{x}_0}\right)_{\underline{x}_1=\underline{a}_1, \underline{x}_0=\underline{a}_0} = 1 \ . \tag{3.94}$$

We say that a function is *unate* in x_i if it is either degenerate, either {0}-degenerate or {1}-degenerate in this variable. A function f is unate in $\underline{x}_0$ if it is unate in each of the variables x_i of $\underline{x}_0$. If the n-variable function f is unate, there exists (at least) one vertex $\underline{a}$ of $\underline{x}$ such that the function f monotonically changes when the n+1 vertices of a path connecting the vertices $\underline{a}$ and $\bar{\underline{a}}$ of the domain of f are runned over. If the function $f(\underline{a}_1,\underline{x}_0)$ is unate we shall say that f is *locally unate* in $\underline{x}_0$ at the vertex $\underline{x}_1=\underline{a}_1$.

Theorem

The function f *is locally unate in* $\underline{x}_0$ *at the vertex* $\underline{x}_1=\underline{a}_1$ *and the transition between the vertices* $\underline{a}=(\underline{a}_1,\underline{a}_0)$ *and* $\underline{b}=(\underline{a}_1,\bar{\underline{a}}_0)$ *is hazard-free if and only if :*

$$\bigvee_{\underline{e}_0} \left[\left(\frac{Sf}{S\underline{x}_0^{\underline{e}_0}} \oplus \frac{\delta f}{\delta \underline{x}_0^{\underline{e}_0}}\right)\right]_{\underline{x}=\underline{a}} = 0, \quad e_i=0,1, \underline{e}_0 \neq \underline{0} \ . \tag{3.95}$$

Proof

The weight of a binary vector is the number of its 1. If the disjunction operation of (3.95) holds first of all for all the vectors of weight 2, the truth of (3.95) implies that all the transitions starting from $\underline{a}$ and for a change of two variables are hazard-free (see the definition of an hazard in section 3.2.8.). If we then re-

quest that the disjunction (3.95) be zero for the vectors $\underline{e}_0$ of weight 3, the transitions involving a change of three of the variables in $\underline{x}_0$ are hazard-free. The condition (3.95) is finally sufficient : this is verified by perfect induction on the number of variables.
If the function f is locally unate in $\underline{x}_0$ at the vertex $\underline{x}_1=\underline{a}_1$ the condition (3.95) is evidently satisfied. □

3.2.15. Theorem

The above considerations lead us to give an interpretation to the function

$$\frac{S^{\underline{A}_0}f}{S\underline{x}_0} \oplus \frac{\delta^{\underline{A}_0}f}{\delta\underline{x}_0} \quad .$$

Consider again the fault detection theory (see also sections 3.1.8 and 3.2.13). A fault is called a *simple fault* if a simple input variable is stuck-at ; it is called a *multiple fault* or a fault of order p if p (p > 1) input variables are stuck-at.
Theorem.

The function

$$\frac{S^{\underline{A}_0}f}{S\underline{x}_0} \oplus \frac{\delta^{\underline{A}_0}f}{\delta\underline{x}_0} \,, \tag{3.96}$$

is a test-function for at least one of the faults of order lower than or equal to p-1 : a (strict) subset of $\underline{x}_0$ *is stuck at* $x_i \oplus x_i^{(A_i)}\ \forall x_i$ *belonging to that subset ; moreover the function (3.96) is not a test-function for the fault of order p :* $\underline{x}_0$ *is stuck-at* $\underline{x}_0 \oplus \underline{x}_0^{(\underline{A}_0)}$.

The usefulness of the above theorem lies in some algorithms for building test-functions. In order to reduce the complexity of computing test functions we consider generally only a well-chosen subset of the possible faults. In this respect it is generally assumed that the occurrence of a fault of order k is more likely than the occurrence of a fault of order p if k < p. However, if one has to detect a fault of order k it may happen that a fault of order p destroys the test-function for that fault of order k. Hence the interest of building test functions for faults of order k < p which are not test- functions for some faults of order p.

3.2.16. Generalization of theorem 3.1.11.

Theorem
The verification of one of the three relations

$$f \oplus \frac{p^{\underline{A}_0}f}{p\underline{x}_0} \equiv 0 \,, \tag{3.97}$$

$$\bar{f} \oplus \frac{q^{\underline{A}_0}\bar{f}}{q\underline{x}_0} \equiv 0 \ , \tag{3.98}$$

$$\frac{p^{\underline{A}_0}f}{p\underline{x}_0} \oplus \frac{q^{\underline{A}_0}f}{q\underline{x}_0} \equiv 0 \ , \tag{3.99}$$

is a necessary and sufficient condition for f *being* $\underline{A}_0$*-degenerate in* $\underline{x}_0$.

Proof

The proof is analogous to that of theorem 3.1.11. Theorem 3.2.16 may also be considered as a corollary of theorem 3.2.6. Indeed, if (3.97) is satisfied, $f \equiv p^{\underline{A}_0}f/p\underline{x}_0$ and f is $\underline{A}_0$-degenerate in $\underline{x}_0$. Moreover if f is $\underline{A}_0$-degenerate in $\underline{x}_0$ it is the disjunction of all its prime implicants which are all $\underline{A}_0$-degenerate in $\underline{x}_0$. Relation (3.98) derives from (3.97) by complementing the two terms of the left-hand side. We may also verify that a function is $\underline{A}_0$-degenerate if and only if its corresponding lower and upper envelopes coincide, i.e. if and only if (3.99) holds. □

Observe also that the condition (3.57) may be expressed in terms of the function $\delta^{\underline{A}_0}f/\delta\underline{x}_0$ i.e. :

$$f \, \frac{\delta^{\underline{A}_0}f}{\delta\underline{x}_0} \equiv 0 \ ; \tag{3.100}$$

Indeed from the inequalities $p^{\underline{A}_0}f/p\underline{x}_0 \leqslant f \leqslant q^{\underline{A}_0}f/q\underline{x}_0$ we successively deduce :

$$\begin{aligned} f \oplus \frac{p^{\underline{A}_0}f}{p\underline{x}_0} &= f\left(\frac{q^{\underline{A}_0}f}{q\underline{x}_0} \oplus \frac{p^{\underline{A}_0}f}{p\underline{x}_0}\right) \ , \\ &= f \, \frac{\delta^{\underline{A}_0}f}{\delta\underline{x}_0} \ . \end{aligned} \tag{3.101}$$

3.2.17. Generalization of theorem 3.1.12.

The following theorem allow us to obtain the expression of the envelopes of f from the Galoisian expansions of this function.

Theorem

The envelopes of f *with respect to* $\underline{x}_0$ *are obtained from the Galoisian expansions of* f *with respect to these variables by replacing the operation* Ⓣ *by* ⊥ .

The proof of this theorem is omitted ; it is similar to that of theorem 3.1.12.

3.2.18. Application of the concept of envelope in switching theory

The importance of the realization of a function f as a disjunction of all its prime implicants or as a conjunction of all its prime implicates has been pointed out by Eichelberger [1965] ; remember first (see section 3.2.8) that according to the function output there are two types of hazards : the *static hazard* and the *dynamic hazard*. Hazards are also classified according to whether they can or cannot be detected by means of tests on the switching function to be realized. *Function hazards* (as they were defined in section 3.2.8) can be detected by performing tests (e.g. the test (3.73)) on the Boolean function while *logic hazards* cannot. These last hazards are thus associated with a switching circuit instead of with the function it realizes ; they are due to stray-delays in the connections and gates which constitute the circuit (Thayse [1971]). The following proposition is due to Eichelberger [1965] .

Proposition

(a) *An AND-OR network has no logic static hazards if its AND-gates are in one-to-one correspondence with the prime implicants of the function being realized.*

(b) *An OR-AND network has no logic static hazards of its OR-gates are in one-to-one correspondence with the prime implicates of the function being realized.*

The importance of the realization of a function as a disjunction of prime lower envelopes or as a conjunction of prime upper envelopes comes from the following proposition (Davio, Deschamps and Thayse [1978])

Proposition

(a) *Consider a three-level OR-AND network realizing f and built up as follows :*
the AND-gates are in one-to-one correspondence with the prime lower envelopes of f ;
the first-level OR-gates, the outputs of which are the inputs of a given AND-gate, are in one-to-one correspondence with the prime implicates of the prime lower envelope realized by that AND-gate.
Then the network thus obtained is :
free of logic static hazards, contains no more logic dynamic hazards than the AND-OR network, the AND gates of which are in one-to-one correspondence with the prime implicants of f.

(b) *Dual statement.*

3.2.19. Continuation of the example 3.1.14.

$$\frac{\Delta f}{\Delta x_0 x_2} = \frac{\Delta f}{\Delta x_1 x_2} = 1 \;;\; \frac{\Delta f}{\Delta x_0 x_1} = \frac{\Delta f}{\Delta x_0 x_1 x_2} = 0 \;;$$

$$\frac{pf}{px_i x_j} = 0 \,,\quad \frac{qf}{qx_i x_j} = 1 \quad \forall\, x_i, x_j \in \{x_0, x_1, x_2\}$$

Expression of f as a disjunction of prime implicants :

$$f = x_0x_2 \vee x_1\bar{x}_2 \vee x_0x_1 \qquad (3.102)$$

Expression of f as a conjunction of prime implicates :

$$f = (x_0 \vee \bar{x}_2)(x_1 \vee x_2)(x_0 \vee x_1) \qquad (3.103)$$

From the Boolean differences we compute the Newton expansions of f ; theorem 3.1.17 allows us to derive the upper envelopes from the Newton expansions.

Computation vertex	Newton expansions	upper envelopes
$(x_2x_1x_0) = (0\ 0\ 0)$	$f = x_0x_2 \oplus x_1 \oplus x_1x_2$	$x_0x_2 \vee x_1$;
$= (0\ 0\ 1)$	$f = x_2 \oplus x_2\bar{x}_0 \oplus x_1 \oplus x_1x_2$	$x_2 \vee x_1$;
$= (0\ 1\ 0)$	$f = 1 \oplus x_2 \oplus x_2x_0 \oplus \bar{x}_1 \oplus \bar{x}_1x_2$	1 ;
$= (1\ 0\ 0)$	$f = x_0 \oplus \bar{x}_2x_0 \oplus \bar{x}_2x_1$	$x_0 \vee x_1\bar{x}_2$;
$= (1\ 0\ 1)$	$f = 1 \oplus \bar{x}_0 \oplus \bar{x}_0\bar{x}_2 \oplus \bar{x}_2 \oplus \bar{x}_2x_1$	1 ;
$= (1\ 1\ 0)$	$f = \bar{x}_2 \oplus x_0 \oplus x_0\bar{x}_2 \oplus \bar{x}_1\bar{x}_2$	$\bar{x}_2 \vee x_0$;
$= (1\ 1\ 1)$	$f = 1 \oplus \bar{x}_0 \oplus \bar{x}_0\bar{x}_2 \oplus \bar{x}_1\bar{x}_2$	1 ;
$= (0\ 1\ 1)$	$f = 1 \oplus x_2\bar{x}_0 \oplus \bar{x}_1 \oplus \bar{x}_1x_2$	1

Expression of f as a conjunction of upper envelopes :

$$f = (x_1 \vee x_0x_2)(x_0 \vee x_1\bar{x}_2) \qquad (3.104)$$

Expression of f as a disjunction of lower envelopes

$$f = x_1(x_0 \vee \bar{x}_2) \vee x_0(x_1 \vee x_2) \qquad (3.105)$$

To the expressions (3.102-3.105) correspond the realizations of figures 3.3(a)-3.3(d) respectively. Let us now interpret these realizations in terms of the presence or absence of some logic hazards.

The function to be realized is $x_0x_2 \vee x_1\bar{x}_2$; it is represented by its truth table 3.III

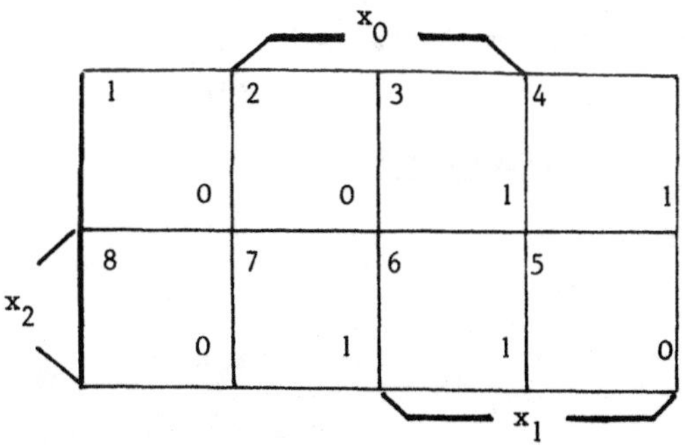

Table 3.III

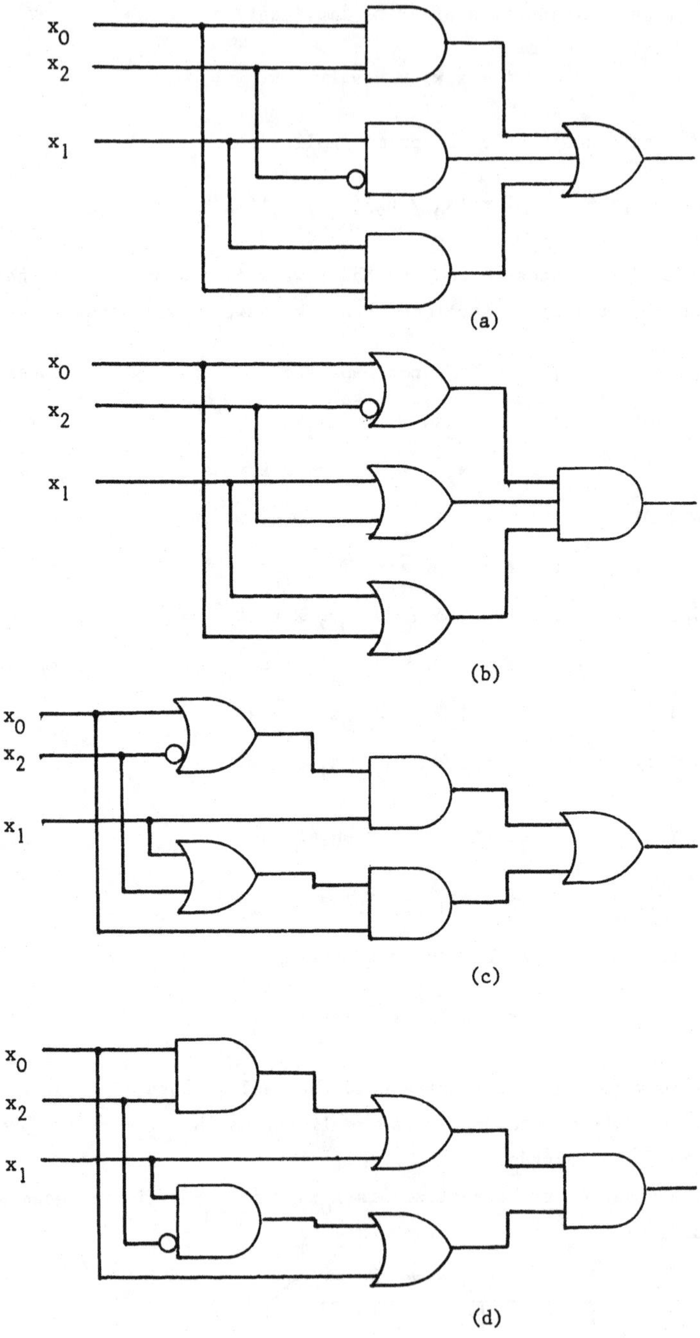

Figure 3.3.

Consider first the two-level realizations of figures 3.3(a) and 3.3(b). Both realizations are free of logic static hazards ; they contain however a lot of transitions experiencing a logic dynamic hazard (see table 3.3).

(a) Dynamic hazard containing transitions for the AND-OR network of Figure 3.3(a) :

$$7 \leftrightarrow 5 \ , \ 2 \leftrightarrow 4 \ , \ 3 \leftrightarrow 5 \ , \ 2 \leftrightarrow 6 \quad .$$

(b) Dynamic hazard containing transitions for the OR-AND network of Figure 3.3(b):

$$7 \leftrightarrow 5 \ , \ 2 \leftrightarrow 4 \ , \ 1 \leftrightarrow 7 \ , \ 4 \leftrightarrow 8 \quad .$$

It is clear, from the above example, that the logic dynamic hazards are produced by function static hazards in the prime implicants or in the prime implicates of the function.

Consider now the three-level realizations of figures 3.3(c) and 3.3(d) ; both realizations are free of logic static hazards and contain dynamic hazards for the transition :

$$2 \leftrightarrow 4 \quad \text{and} \quad 7 \leftrightarrow 5 \quad .$$

From the Boolean differences we shall now compute the sensitivities $Sf/S\underline{x}_0$, the δ-functions $\delta f/\delta\underline{x}_0$ and hence the function $\delta f/\delta\underline{x}_0 \oplus Sf/S\underline{x}_0$; the computations are gathered in Table 3.IV.

$\underline{x}_0$	$Sf/S\underline{x}_0$	$\delta f/\delta\underline{x}_0$	$Sf/S\underline{x}_0 \oplus \delta f/\delta\underline{x}_0$
x_0x_1	1	1	0
x_0x_2	$\bar{x}_0 \oplus x_1 \oplus x_2$	1	$x_0 \oplus x_1 \oplus x_2$
x_1x_2	$x_0 \oplus x_1 \oplus x_2$	1	$\bar{x}_0 \oplus x_1 \oplus x_2$
$x_0x_1x_2$	$\bar{x}_0 \oplus x_1$	1	$x_0 \oplus x_1$

Table 3.IV

Any network realizing the switching function $f=x_0x_2 \vee x_1\bar{x}_2$ experiences a function static hazard for the following transitions :

Changing variables	;	equation characterizing the initial or final variables value.
x_0,x_2	;	$x_0 \oplus x_1 \oplus x_2 = 1$
x_1,x_2	;	$\bar{x}_0 \oplus x_1 \oplus x_2 = 1$
x_0,x_1,x_2	;	$x_0 \oplus x_1 = 1$.

4. Application to switching theory

4.1. Introduction.

The essential of chapter 3 has been devoted to proving a set of theorems and formulas. In the present chapter 4 we apply these theorems and formulas to solve problems occurring in switching theory. As quoted in the introduction, problems are solved by means of algorithms or computation schemata. Observe that a single formula may induce several algorithms for solving a given problem. This will appear clearly in the sequel of this chapter. We shall also restrict ourselves to the solution of problems occurring in combinatorial networks : the study of sequential networks will thus be excluded.

We consider three main types of problems, i.e. the synthesis problems, the analysis problems and the problems related to the detection of some particular functional properties. This last kind of problem may also be considered either as an analysis problem or as a synthesis problem : to detect functional properties may be useful at the same time to the analysis and to the synthesis of switching networks. In the course of this chapter we exhaustively deal with examples;the approach to the proposed algorithms will thus be simplified.

In section 4.2 we deal with universal algorithms. We show e.g. how the extended state vector $\underline{\phi}^{(\dagger)}(f)$ (with $\dagger=\vee,\wedge,\oplus$ or $\odot$) may be used in deriving functional properties of f ; in section 4.3 we deal with particular algorithms : we show e.g. how the extended state vectors $\underline{\phi}^{(\vee)}(f)$, $\underline{\phi}^{(\wedge)}(f)$ and $\underline{\phi}^{(\oplus)}(f)$ may be used in deriving functional properties of f attached to the extended state vector considered.

4.2. Universal algorithms

4.2.1. Algorithms grounded on the extended vector

4.2.1.1. Formulas and theorems

The extended state vector of a Boolean function $f(\underline{x}),\underline{x}=(x_{n-1},\dots,x_1,x_0)$, was defined in section 2.3.2 ; we recall here its fundamental properties.

The *partial extended vector* with respect to the variable x_{n-1} and according to the law $\dagger$ is denoted $\underline{\phi}_{x_{n-1}}^{(\dagger)}(f)$ and is defined as follows :

$$\underline{\phi}_{x_{n-1}}^{(\dagger)}(f) = [\, f_{x_{n-1}=0}\, ,\, f_{x_{n-1}=1}\, ,\, f_{x_{n-1}=0} \dagger f_{x_{n-1}=1}\,] \tag{4.1}$$

The *extended vector* $\underline{\phi}_{\underline{xy}}^{(\dagger)}(f)$ of an (n+1)-variable function $f(\underline{xy})$ is defined by using the induction formula :

$$\underline{\phi}_{\underline{xy}}^{(\dagger)}(f)= [\,\underline{\phi}_{\underline{x}}^{(\dagger)}(f(\underline{x},0)),\ \underline{\phi}_{\underline{x}}^{(\dagger)}(f(\underline{x},1)),\phi_{\underline{x}}^{(\dagger)}(f(\underline{x},0)\ \dagger\ f(\underline{x},1))] \tag{4.2}$$

The extended vector of an n-variable function is a (1×3^n) matrix ; it may be obtained by matrix multiplication from the truth vector (see (2.43)).

The usefulness of the extended vector is due to the fact that it contains, in a relatively condensed form, the truth vectors of f and of all its differences $\uparrow f/\uparrow x_i$, $\uparrow f/\uparrow x_i x_j$,... $\uparrow f/\uparrow \underline{x}$ (any one of these differences will be denoted $\uparrow f/\uparrow \underline{x}^{\underline{e}}$, $\underline{e}=(e_{n-1},\ldots,e_1,e_0)$, $e_i \in \{0,1\}\ \forall i$). A rough proof of this fact may be stated as follows : observe that $\underline{\phi}^{(\uparrow)}_{x_{n-1}}(f)$ contains the function f and its differences $\uparrow f/\uparrow x_{n-1}$; the induction formula (4.2) allows us to state that $\underline{\phi}^{(\uparrow)}_{x_{n-1}x_{n-2}}(f)$ contains the function f and its differences $\uparrow f/\uparrow x_{n-1}$, $\uparrow f/\uparrow x_{n-2}$ and $\uparrow f/\uparrow x_{n-1}x_{n-2}$. Finally an inductive argument allows us to state that $\phi^{(\uparrow)}_{\underline{x}}$ contains f and all its differences $\uparrow f/\uparrow \underline{x}^{\underline{e}}$ (remember that the exponentiation x^e is defined as $x^e=x$ if e=1 and $x^e=1$ if e=0).

Observe that the extended vector does not provide us with a direct information about the differences of the form :

$$\frac{\uparrow^{A_0A_1\ldots A_{p-1}}f}{\uparrow x_0x_1\ldots x_{p-1}} \quad , \quad A_i \neq \{0,1\} \ . \tag{4.3}$$

This drawback may be circumvented by observing that differences of the from (4.3) may always be expressed in terms of differences of the form $\uparrow f/\uparrow \underline{x}^{\underline{e}}$. Let $(\underline{x}_1,\underline{x}_0)$ be a partition of $\underline{x}$ with $\underline{x}_0 = (x_{p-1},\ldots,x_1,x_0)$ and $\underline{x}_1=(x_{n-1},\ldots,x_p)$; the following formulas have been proved in section 3.2.2. For $\uparrow$ the disjunction $\vee$, the difference (4.3), generally denoted $\uparrow^{\underline{A}_0}f/\uparrow \underline{x}_0$, may be expressed as :

$$\frac{q^{\underline{A}_0}f}{q\underline{x}_0} = \bigvee_{\underline{e}_0}\Big[\bigwedge_{i=0,p-1}(x_i^{(A_i)})^{e_i} \wedge \frac{qf}{\underline{x}_0^{\underline{e}_0}}\Big] \ , \ \underline{e}_0=(e_0,e_1,\ldots,e_{p-1}),e_i \in\{0,1\}\forall i \ . \tag{4.4}$$

For $\uparrow$ the conjunction $\wedge$, the difference (4.3) may be expressed as :

$$\frac{p^{\underline{A}_0}f}{p\underline{x}_0} = \bigwedge_{\underline{e}_0}\Big[\bigvee_{i=0,p-1}\overline{(x_i^{(A_i)})^{e_i}} \vee \frac{pf}{p\underline{x}_0^{\underline{e}_0}}\Big] \ . \tag{4.5}$$

If $\top$ means either the disjunction $\vee$ or the conjunction $\wedge$, denote by $\perp$ the operation dual to $\top$ and by $\textcircled{T}$ the Boolean ring law which is distributed by $\top$: if $\top$ is the disjunction , then $\textcircled{T}$ is the identity $\odot$ while if $\top$ is the conjunction, then $\textcircled{T}$ is the modulo-2 sum $\oplus$; these notations allow us to express the formulas (4.4,4.5) in the following unique form :

$$\frac{T^{\underline{A}_0} f}{T\underline{x}_0} = \underset{\underline{e}_0}{T} \left[\underset{i=0,p-1}{\bot} \left((x_i^{(A_i)})^{e_i} \, ⓣ \, 1 \right) \bot \frac{Tf}{T\underline{x}_0^{\underline{e}_0}} \right], \; T=\vee \text{ or } \wedge. \quad (4.6)$$

We obtained also the following expression for the differences $ⓣ^{\underline{A}_0} f / ⓣ\underline{x}_0$ (see (3.57))

$$\frac{ⓣ^{\underline{A}_0} f}{ⓣ\,\underline{x}_0} = \underset{i=0,p-1}{T} (x_i^{(A_i)} \, ⓣ \, 0) \; T \; \frac{ⓣ f}{ⓣ\underline{x}_0}, \; ⓣ = \odot \text{ or } \oplus . \quad (4.7)$$

In summary the expressions (4.6,4.7) allow us to obtain the differences $\uparrow^{\underline{A}_0} f/\uparrow\underline{x}_0$ in terms of the differences $\uparrow f/\uparrow\underline{x}_0^{\underline{e}_0}$ which are themselves contained in the extended vector $\underline{\phi}^{(\uparrow)}(f)$; remember also that the main usefulness of these differences is to obtain e.g. prime implicants, prime implicates, test functions, series expansions ... etc. of f. The two theorems below provide us with a direct way for obtaining from the extended vector, prime implicants, prime implicates, ring-sum expansions and Boolean differences of f.

Denote by $[AM]_M$ a matrix multiplication with additive law A and multiplicative law M and by $\otimes$ the Kronecker matrix product with multiplicative law M ; let also e_T be the zero element for the law T.

Theorem 2.3.3.

$$f = \underline{\phi}^{(T)}(f) \; [\bot \; T] \left(\overset{T}{\underset{i=n-1,0}{\otimes}} \begin{bmatrix} x_i \, ⓣ \, 1 \\ x_i \, ⓣ \, 0 \\ e_T \end{bmatrix} \right) \quad (4.8)$$

If T is the conjunction the matrix product (4.8) gives us an expression of f as a conjunction of all its implicates ; if T is the disjunction the matrix product (4.8) gives us an expression of f as the disjunction of all its implicants.

Theorem 2.3.5.

The 2^n Galois expansions of $f(\underline{x})$ are obtained from the matrix relation :

$$f = \underline{\phi}^{(ⓣ)}(f) \; [ⓣ \; T] \left(\overset{T}{\underset{i=n-1,0}{\otimes}} \begin{bmatrix} h_i \, ⓣ \, 1 \\ h_i \, ⓣ \, 0 \\ x_i \, ⓣ \, h_i \end{bmatrix} \right) \quad (4.9)$$

by giving to the vector $\underline{h}=(h_{n-1},\ldots,h_1,h_0)$ *its* 2^n *possible values ; the classical Reed-Muller expansions are the Galois expansions with* $\textcircled{T} = \oplus$.

4.2.1.2. <u>The algebra derived from the law ↑</u>

We observe that the extended vector $\underline{\phi}^{(\uparrow)}(f)$ contains in a condensed form an information about the most important concepts in switching theory. It is always straightforward to compute the extended vector $\underline{\phi}^{(\uparrow)}(f)$ of a Boolean function f when the law "↑" has been chosen : we have then to apply the usual rules of the Boolean algebra. Assume now that the function f is given by its truth vector and that we want to evaluate an expression of the extended vector without resorting to a fixed choice of the law ↑ .

Observe first (in view of formulas (4.1,4.2)) that the 0 and 1 present in the extended vector come either from the 0 and 1 present in the truth vector, or from a composition (with respect to the law ↑) of an even number of 1 and of 0. In the extended vector, and for any law ↑, we have thus to be able to recognize the presence of 0,1 but also of composition of 0's (written $\underline{0}$), of 1's (written $\underline{1}$), of 0's with an even number of 1's (written e) and of 0's with an odd number of 1's (written ∅). Composition of 0 and 1 with respect to the law ↑ may thus be depicted by the following truth tables 4.I(a), 4.I(b).

↑	0	1
0	$\underline{0}$	∅
1	∅	$\underline{1}$

(a)

↑	$\underline{0}$	$\underline{1}$	e	∅
$\underline{0}$	$\underline{0}$	e	e	∅
$\underline{1}$	e	$\underline{1}$	e	∅
e	e	e	e	∅
∅	∅	∅	∅	e

(b)

<u>Table 4.I</u>

We obtain an extended vector $\underline{\phi}(f)$ of 3^n elements, each of the elements belonging to the set $\{0,1,\underline{0},\underline{1},e,\emptyset\}$; from theorems 2.3.3 and 2.3.5 deduce that :

- the implicants of f correspond to the entries 1 and $\underline{1}$ of $\underline{\phi}(f)$;
- the implicates of f correspond to the entries 0 and $\underline{0}$ of $\underline{\phi}(f)$;
- the elements of the Reed-Muller expansions of f correspond to the entries 1 and ∅ of $\underline{\phi}(f)$;
- the elements of the dual Galois expansions of f correspond to the entries 0 and ∅ of $\underline{\phi}(f)$.

4.2.1.3. <u>Continuation of the example 3.1.14.</u>

Consider the three-variable function given by its truth vector :

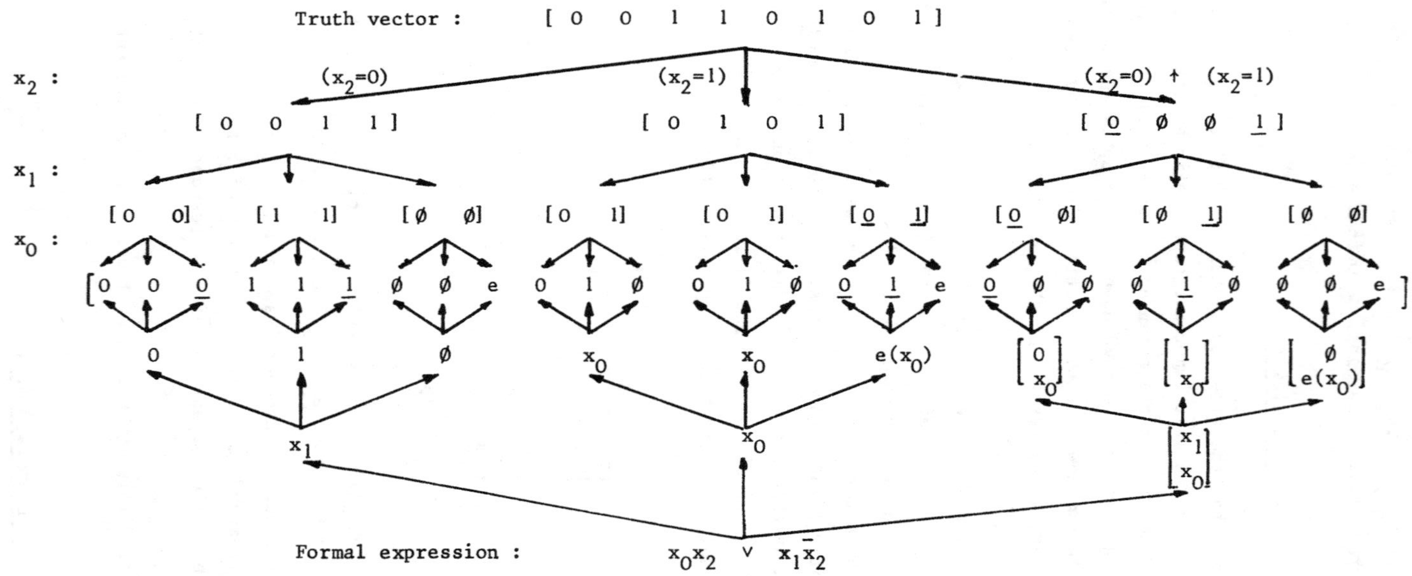
Truth vector : [0 0 1 1 0 1 0 1]
x2 :
(x2=0)
(x2=1)
(x2=0) + (x2=1)
[0 0 1 1]
[0 1 0 1]
[0 Ø Ø 1]
x1 :
[0 0] [1 1] [Ø Ø] [0 1] [0 1] [0 1] [0 Ø] [Ø 1] [Ø Ø]
x0 :
0 0 0 1 1 1 Ø Ø e 0 1 Ø 0 1 Ø 0 1 e 0 Ø Ø Ø 1 Ø Ø Ø e
0 1 Ø x0 x0 e(x0)
x1
x0
Formal expression : x0x2 v x1x̄2

Figure 4.1.

	Dual Galois expansions : $\emptyset$,0	Reed-Muller expansions : $\emptyset$,1	Implicates, $\underline{0}$, 0	Implicants, 1, $\underline{1}$
0	$h_2 \vee h_1 \vee h_0$		$x_0 \vee x_1 \vee x_2$	
0	$h_2 \vee \bar{h}_1$		$\bar{x}_0 \vee x_1 \vee x_2$	
$\underline{0}$			$x_1 \vee x_2$	
1		$\bar{h}_2 h_1 \bar{h}_0$		$\bar{x}_0 x_1 \bar{x}_2$
1		$h_2 h_1 h_0$		$x_0 x_1 \bar{x}_2$
$\underline{1}$				$x_1 \bar{x}_2$
$\emptyset$	$h_2 \vee h_0 \vee \bar{\dot{x}}_1$	$\bar{h}_2 \bar{h}_0 \dot{x}_1$		
$\emptyset$	$h_2 \vee \bar{h}_0 \vee \bar{x}_1$	$h_2 h_0 \dot{x}_1$		
e				
0	$\bar{h}_2 \vee h_1 \vee h_0$		$x_0 \vee x_1 \vee \bar{x}_2$	
1		$h_2 \bar{h}_1 h_0$		$x_0 \bar{x}_1 x_2$
$\emptyset$	$\bar{h}_2 \vee h_1 \vee \bar{\dot{x}}_0$	$h_2 \bar{h}_1 \dot{x}_0$		
0	$\bar{h}_2 \vee \bar{h}_1 \vee h_0$		$x_0 \vee \bar{x}_1 \vee \bar{x}_2$	
1		$h_2 h_1 h_0$		$x_0 x_1 x_2$
$\emptyset$	$\bar{h}_2 \vee \bar{h}_1 \vee \bar{\dot{x}}_0$	$h_2 h_1 \dot{x}_0$		
$\underline{0}$			$x_0 \vee \bar{x}_2$	
$\underline{1}$				$x_0 x_2$
e				
$\underline{0}$			$x_0 \vee x_1$	
$\emptyset$	$h_1 \vee \bar{h}_0 \vee \bar{\dot{x}}_2$	$\bar{h}_1 h_0 \dot{x}_2$		
$\emptyset$	$h_1 \vee \bar{\dot{x}}_2 \vee \bar{\dot{x}}_0$	$\bar{h}_1 \dot{x}_2 \dot{x}_0$		
$\emptyset$	$\bar{h}_1 \vee h_0 \vee \bar{x}_2$	$h_1 \bar{h}_0 x_2$		
$\underline{1}$				$x_0 x_1$
$\emptyset$	$\bar{h}_1 \vee \bar{\dot{x}}_2 \vee \bar{\dot{x}}_0$	$h_1 \dot{x}_2 \dot{x}_0$		
$\emptyset$	$h_0 \vee \bar{\dot{x}}_2 \vee \bar{\dot{x}}_1$	$\bar{h}_0 \dot{x}_2 \dot{x}_1$		
$\emptyset$	$\bar{h}_0 \vee \bar{\dot{x}}_2 \vee \bar{\dot{x}}_1$	$h_0 \dot{x}_2 \dot{x}_1$		
e				

Table 4.II.

$$[\;0\;\;0\;\;1\;\;1\;\;0\;\;1\;\;0\;\;1\;]$$

In Figure 4.1 we show how the extended vector is obtained from the truth vector by means of three iterations of the type (4.1) with respect to the variables x_2, x_1 and x_0 respectively.

The corresponding implicants, implicates, Reed-Muller terms and dual Galois terms are written down in Table 4.II ; for the sake of conciseness the literal $x_i^{(h_i)}$ is written $\dot{x}_i$. One deduces e.g. that the Reed-Muller expansion and the dual Galois expansion at $(h_2h_1h_0)=(0\;0\;0)$ are respectively :

$$x_1 \oplus x_0x_2 \oplus x_1x_2$$

$$\bar{x}_0 \odot (\bar{x}_0 \vee \bar{x}_2) \odot (\bar{x}_2 \vee \bar{x}_1) \odot 0$$

Observe that the Boolean differences, which are implicitely contained in the Reed-Muller terms may also directly be obtained from the extended vector by making use of the theorem below which constitutes an immediate generalization of theorem 2.3.3.

Theorem

The Galois expansions of $f(\underline{x})$ *and of its* (2^n-1) *differences* $\textcircled{T}f/\textcircled{T}\underline{x}^{\underline{e}}$, $e_i \in \{0,1\}$, $\underline{e} \neq (0,0,\ldots 0)$ *are obtained from the matrix relation :*

$$\underline{\phi}^{(\textcircled{T})}(f)\;[\textcircled{T}\,T\,]\;\Big(\bigotimes_{i=n-1,0}^{T}\begin{bmatrix} h_i \textcircled{T} 1 & e_\perp \\ h_i \textcircled{T} 0 & e_\perp \\ x_i \textcircled{T} h_i & e_\top \end{bmatrix}\Big) \qquad (4.10)$$

Observe finally that the extended vector may also be obtained from any formal expression giving f. In this case the composition rules, as defined by Table 4.I(a,b), are no longer sufficient to evaluate the extended vector $\underline{\phi}^{(\uparrow)}(f)$; we shall add the following rules :

- for any expression $g(\underline{x})$, the difference $g(x=0) \uparrow g(x=1)$ will be written as a (2×1) matrix, i.e. :

$$g(x=0) \uparrow g(x=1) = \begin{bmatrix} g(x=0) \\ g(x=1) \end{bmatrix},$$

the ordering of the two elements in this matrix being unessential.

- an iterative use of the above rule leads to the generation of $(2p\times 1)$ matrices ; the number of rows of these matrices may be reduced by observing that the generation of the extended vectors needs only the knowledge of the different expressions present in the matrix and of their parity. In this respect the presence of an odd number of expressions g will be written $\emptyset(g)$ while the presence of an even number of expressions g will be written $e(g)$. As above, $\emptyset(1)$ and $e(1)$ will be written $\emptyset$ and e respectively.

Example (Continued)

In Figure 4.1. we show how the extended vector is obtained from the expression $x_0x_2 \vee x_1\bar{x}_2$ of f.

Remarks

1. Assume that two functions f and g be given either by their truth vector, or by their extended vector ; the composition $f \uparrow g$ of these two functions with respect to the law $\uparrow$ is then obtained by the composition rules of tables 4.I(a), I(b).

2. In many applications we are interested in the extended vector $\underline{\phi}_{\underline{x}}^{(T)}(f)$ where T is either the conjunction or the disjunction ; in this case we have to be able to recognize the presence in the extended vectors of either only 0's, either 1's or both 0's and 1's. In this respect the elements of the extended vector are 0,1 and $\emptyset$; they are either deduced from the Table 4.III :

T	0	1	∅
0	0	∅	∅
1	∅	1	∅
∅	∅	∅	∅

Table 4.III.

or from the extended vector $\phi_{\underline{x}}^{(\uparrow)}(f)$ by changing the elements as follows :

$$\underline{0} \rightarrow 0, \ \underline{1} \rightarrow 1, \ e \rightarrow \emptyset \quad .$$

4.2.2. Algorithms grounded on the generalized consensus

4.2.2.1. Introduction

A well-known procedure for finding prime implicants of a switching function is based on what Quine first called the *consensus* of implicants. The original method developed by Quine is generally referred to as the *iterative consensus* ; it is fully detailed in most of the books dealing with switching theory (see e.g., Mc Cluskey [1965]).

This method has been improved by Tison [1969,1967,1971] who suggested a most efficient algorithm which is called the *generalized consensus*. Besides the original papers by Tison, a description of this algorithm may also be found in the books by Kuntzmann [1965] and Kuntzmann and Naslin [1967] . The author (Thayse [1978]) showed that the concepts of meet and join differences are a convenient mathematical support for both the iterative and the generalized consensus. The purpose of this section is to show that the use of the differences $\uparrow f/\uparrow x$ (instead of the meet and join differences) together with the operations defined in tables 4.I and 4.III allow us to apply the consensus algorithms to a broader class of problems than those defined by Tison.

Consider a Boolean function of e.g. three variables and introduce f and its differences $\uparrow f/\uparrow \underline{x}^e$ in the scheme of figure 4.2(a). The extended vector algorithm for this function may be schematized as follows (see section 4.2.1 and figure 4.2(b))

Steps :

(a) from f deduce $\uparrow f/\uparrow x_2$;

(b) from f and $\uparrow f/\uparrow x_2$ deduce $\uparrow f/\uparrow x_1$ and $\uparrow f/\uparrow x_2 x_1$ respectively ;

(c) from f, $\uparrow f/\uparrow x_2$, $\uparrow f/\uparrow x_1$ and $\uparrow f/\uparrow x_2 x_1$ deduce $\uparrow f/\uparrow x_0$, $\uparrow f/\uparrow x_2 x_0$, $\uparrow f/\uparrow x_1 x_0$ and $\uparrow f/\uparrow x_2 x_1 x_0$ respectively.

The iterative consensus, as it was originally developed by Tison, may be interpreted as follows in terms of the differences $\uparrow f/\uparrow \underline{x}^e$

Steps :

(a) from f deduce $\uparrow f/\uparrow x_0$, $\uparrow f/\uparrow x_1$ and $\uparrow f/\uparrow x_2$;

(b) from $\uparrow f/\uparrow x_0$ deduce $\uparrow f/\uparrow x_0 x_1$ and $\uparrow f/\uparrow x_0 x_2$ and from $\uparrow f/\uparrow x_1$ deduce $\uparrow f/\uparrow x_1 x_2$;

(c) from $\uparrow f/\uparrow x_0 x_1$ deduce $\uparrow f/\uparrow x_0 x_1 x_2$.

For an n-variable function, the iterative consensus requests to perform the difference operations $\uparrow f/\uparrow \underline{x}^e$ (2^n-1) times (see also figures 4.2(c)).

The generalized consensus was interpreted in terms of meet differences by the author (Thayse [1978]) :

Steps :

(a) from f, deduce $f \vee pf/px_2 = f_0$;

(b) from f_0, deduce $f_0 \vee pf_0/px_1 = f_1$;

(c) from f_1, deduce $f_1 \vee pf_1/px_0 = f_2$.

We show that, starting from an irredundant disjunction of cubes of f, the generalized consensus gives us after the application of n difference operations (and the use of the distributive and absorption laws) an expression f_n which is the disjunction of all the prime implicants of f. Starting with a function f given as a disjunction of cubes, we show that if at each step of this algorithm we delete in the expression $f_i \vee pf_i/px_j$ the cubes strictly contained in other cubes, the remaining cubes are exactly those obtained at the corresponding step of the generalized consensus algorithm. These steps are schematically described in figure 4.2(d) for a three-variable function.

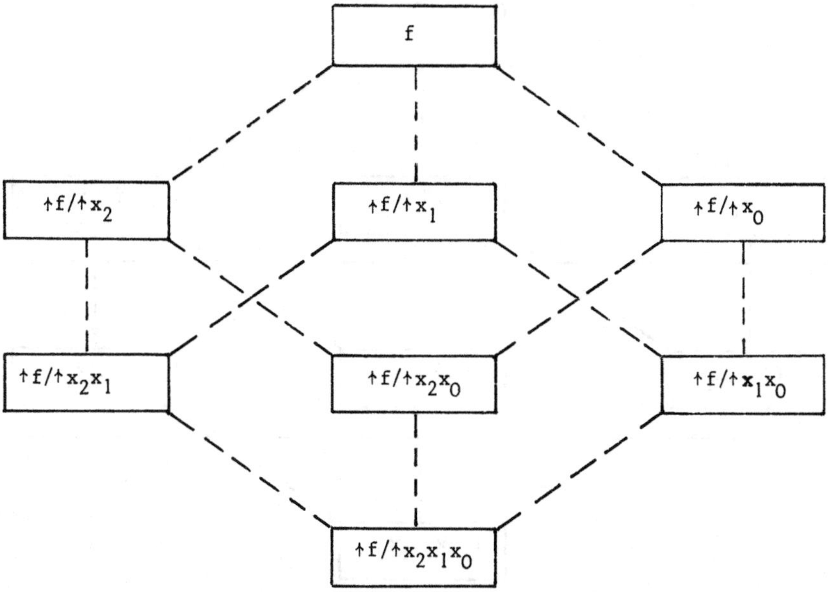
f
↑f/↑x2
↑f/↑x1
↑f/↑x0
↑f/↑x2x1
↑f/↑x2x0
↑f/↑x1x0
↑f/↑x2x1x0

Figure 4.2. (a)
(b)

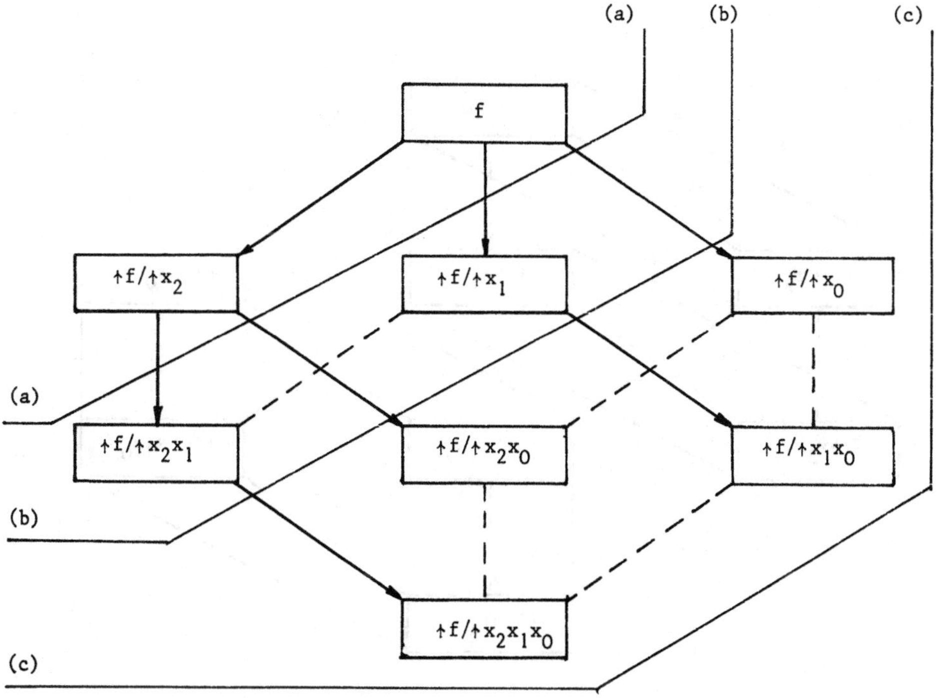
(a)
(b)
(c)
f
↑f/↑x2
↑f/↑x1
↑f/↑x0
↑f/↑x2x1
↑f/↑x2x0
↑f/↑x1x0
↑f/↑x2x1x0
(a)
(b)
(c)

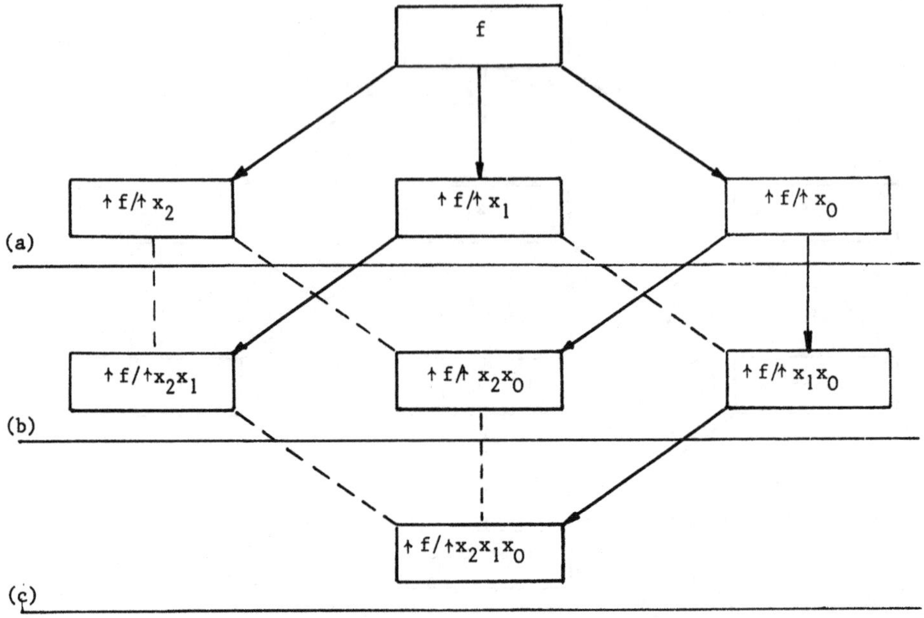
f
$\uparrow f/\uparrow x_2$
$\uparrow f/\uparrow x_1$
$\uparrow f/\uparrow x_0$
(a)
$\uparrow f/\uparrow x_2x_1$
$\uparrow f/\uparrow x_2x_0$
$\uparrow f/\uparrow x_1x_0$
(b)
$\uparrow f/\uparrow x_2x_1x_0$
(c)

Figure 4.2. (c)

(d)

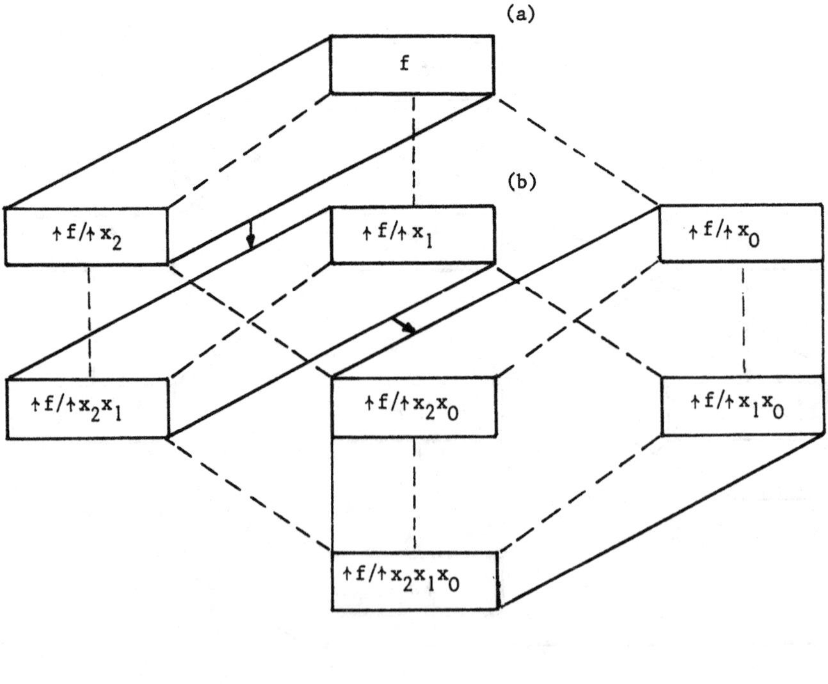
(a)
f
(b)
$\uparrow f/\uparrow x_2$
$\uparrow f/\uparrow x_1$
$\uparrow f/\uparrow x_0$
$\uparrow f/\uparrow x_2x_1$
$\uparrow f/\uparrow x_2x_0$
$\uparrow f/\uparrow x_1x_0$
$\uparrow f/\uparrow x_2x_1x_0$

4.2.2.2. The generalized consensus with respect to the law $\uparrow$

Let T be either the disjunction $\vee$ or the conjunction $\wedge$ and $\perp$ be the dual of these operations ; the generalized consensus algorithm may be interpreted as follows :

Put $f=f_{-1}$;

Perform successively the operations :

$$f_i = f_{i-1} \perp (Tf_{i-1}/Tx_{n-i-1}) \tag{4.11}$$

for the successive values of the index : $i=0,1,\ldots,n-1$.

If T represents the conjunction the application of (4.11) gives us f as the disjunction of its prime implicants while if T represents the disjunction the application of (4.11) gives us the conjunction of its prime implicates. Observe that the use of the laws $T,\perp$ and of the corresponding distributive and absorption laws leads to more cumbersome operations than a separate treatment of the cases $T=\vee$ and $T=\wedge$. This is generally true when f is given by a formal expression ; the purpose of this section is to show that the general consensus algorithm, as expressed by the iterative process (4.11) together with the laws described in tables 4.I and 4.III provide us with simple computation methods when f is given by its truth table.

As suggested in section 4.2.1.3. let us first write the difference Tf/Tx in a matrix form, i.e. :

$$\frac{Tf}{Tx} = \begin{bmatrix} f(x=0) \\ f(x=1) \end{bmatrix}$$

We have :

$$f \perp \frac{Tf}{Tx} = f \perp \begin{bmatrix} f(x=0) \\ f(x=1) \end{bmatrix}$$

$$= \begin{bmatrix} f \perp f(x=0) \\ f \perp f(x=1) \end{bmatrix} \tag{4.12}$$

$$= \begin{bmatrix} f(x=0) & \perp f/\perp x \\ \perp f/\perp x & f(x=1) \end{bmatrix} \tag{4.13}$$

The rows of the (2×2) matrix (4.13) are the partial truth vectors with respect to x of the elements of the (2×1) matrix (4.12) respectively. This suggests us to define a generalized consensus operation as follows .

Let $\uparrow$ be a binary operation ;

Step 1 The consensus of f with respect to x_{n-1} is denoted $\underline{\psi}^{(\uparrow)}_{x_{n-1}}(f)$ and is defined as follows :

$$\underline{\psi}^{(\uparrow)}_{x_{n-1}}(f) = \left[\begin{array}{c|c} f(x_{n-1}=0) & \uparrow f/\uparrow x_{n-1} \\ \hline \uparrow f/\uparrow x_{n-1} & f(x_{n-1}=1) \end{array}\right] \tag{4.14}$$

Step 2

$$\underline{\psi}^{(\uparrow)}_{x_{n-1}x_{n-2}}(f) = \underline{\psi}^{(\uparrow)}_{x_{n-2}}(\underline{\psi}^{(\uparrow)}_{x_{n-1}}(f)) = \left[\begin{array}{c|c} \underline{\psi}^{(\uparrow)}_{x_{n-2}}(f(x_{n-1}=0)) & \underline{\psi}^{(\uparrow)}_{x_{n-2}}(\uparrow f/\uparrow x_{n-1}) \\ \hline \underline{\psi}^{(\uparrow)}_{x_{n-2}}(\uparrow f/\uparrow x_{n-1}) & \underline{\psi}^{(\uparrow)}_{x_{n-2}}(f(x_{n-1}=1)) \end{array}\right]$$

Step n The induction process initialized by steps 1 and 2 produces at the (n-1)-th step a $(2^{n-1}\times 2^{n-1})$ matrix $[a_{ij}(x_0)]$; $\underline{\psi}^{(\uparrow)}_{\underline{x}}(f)$ is the $(2^n\times 2^n)$ matrix :

$$\underline{\psi}^{(\uparrow)}(f) = [\underline{\psi}^{\uparrow}_{x_0}(a_{ij})]$$

Theorem

The rows of the $(2^n\times 2^n)$ *matrix*

$$\underline{\psi}^{(\uparrow)}(f)$$

are :

- *either the truth vectors of the* 2^n *upper envelopes of* f *for* $\uparrow = \vee$,
- *either the truth vectors of the* 2^n *lower envelopes of* f *for* $\uparrow = \wedge$,
- *either the* $\underline{h}$*-transforms of the* 2^n *Reed-Muller expansions of* f *for* $\uparrow = \oplus$,
- *or the* $\bar{\underline{h}}$*-transforms of the* 2^n *dual Galois expansions of* f *for* $\uparrow = \odot$.

Proof

The upper and lower envelopes of f are obtained (see section 3.1.12) by n applications of the operator $T^{\{\bar{h}_i\}}/Tx_i$, i=0,1,...,n-1 to the function f ; recall that :

$$\frac{T^{\{\bar{h}_i\}}f}{Tx_i} = f(h_i)\ T\,[\,(x_i \oplus h_i) \perp \frac{Tf}{Tx_i}\,] \tag{4.15}$$

The Reed-Muller and dual Galois expansions of f are obtained (see section 3.1.11) by n applications of the operator $G^{\{h_i\}}f/G(x_i)$, i=0,1,...,n-1 to the function f ; it is defined as :

$$\frac{G^{\{h_i\}} f}{G(x_i)} = f(h_i) \circledT [(x_i \circledT h_i) \; T \; \frac{\circledT f}{\circledT x_i}] \, . \tag{4.16}$$

The proof derives from a comparison of (4.13), (4.15) and (4.16) ; from (4.13) we deduce that the two rows of $\underline{\psi}^{\uparrow}_{x_{n-1}}(f)$ are formed by the elements $f(h_{n-1})$ and $\uparrow f/\uparrow x_{n-1}$, i.e. :

$$\underline{\psi}^{\uparrow}_{x_{n-1}}(f) = \begin{bmatrix} f(x_{n-1}=0) & \uparrow f/\uparrow x_{n-1} \\ \uparrow f/\uparrow x_{n-1} & f(x_{n-1}=1) \end{bmatrix}$$

An elementary induction process allows us to state that :

- the rows of $\underline{\psi}^{T}(f)$ are the truth vectors of the envelopes of f ;
- the rows of $\underline{\psi}^{\circledT}(f)$ are the $\underline{h}$-transforms of f, with $\underline{h}=(h_{n-1},\ldots,h_1,h_0)$ □

<u>Remark</u> : <u>The $\underline{h}$-transform of a Boolean function f</u> (see also Davio, Deschamps and Thayse [1978] ,p. 265).

Consider a switching function $f(\underline{x})$; the truth vector of that function is denoted by $[f_{\underline{e}}]$; we also consider the $\underline{h}$-Reed-Muller expansion of $f(\underline{x})$, i.e. the Reed-Muller expansion of $f(\underline{x})$ at $\underline{x}=\underline{h}$. The coefficients of the $\underline{h}$-Reed-Muller expansion expressed in the base (called $\underline{h}$-Reed-Muller or $\underline{h}$-R-M base) namely :

$$\mathop{\hat{\otimes}}_{i=n-1,0} \begin{bmatrix} x_i^{h_i} \\ \\ x_i^{\bar{h}_i} \end{bmatrix} \tag{4.17}$$

form the vector $[R\text{-}Mf_{\underline{h}}]$ called $\underline{h}$-Reed-Muller vector (and also for sake of conciseness $\underline{h}$-R-M vector) of $f(\underline{x})$.

One has :

$$\begin{bmatrix} \bar{x}_i^{h_i} \\ x_i^{\bar{h}_i} \end{bmatrix} = \begin{bmatrix} 1 \\ x_i \end{bmatrix} \text{ iff } h_i=0 \, ,$$
$$= \begin{bmatrix} \bar{x}_i \\ 1 \end{bmatrix} \text{ iff } h_i=1 \, . \tag{4.18}$$

The $\underline{h}$-transform of $f(\underline{x})$ is the switching function $f^{\underline{h}}(\underline{x})$ defined by :

$$[f^{\underline{h}}_{\underline{e}}] = [R\text{-}Mf_{\underline{h}}] \tag{4.19}$$

or equivalently by :

$$[\ \text{R-M}\ f^{\underline{h}}_{\underline{h}}\ ] = [\ f_{\underline{e}}\ ] \tag{4.20}$$

The above relations simply express the fact that the functions $f(\underline{x})$ and $f^{\underline{h}}(\underline{x})$ have interchanged truth and $\underline{h}$-R-M vectors, and it turns out that the computation of the $\underline{h}$-R-M vectors of $f(\underline{x})$ may be replaced by that of the truth vector of its $\underline{h}$-R-M transform $f^{\underline{h}}(\underline{x})$.

From (4.17-4.20) we deduce that the weight (i.e. the number of terms) of the Reed-Muller expansion at $\underline{x}=\underline{h}$ is the number of 1 in the truth vector of $f^{\underline{h}}(\underline{x})$. From (4.17,4.18) we deduce the Reed-Muller expansion at $\underline{x}=\underline{h}$ from the truth vector of $f^{\underline{h}}(\underline{x})$.

4.2.2.3. Continuation of example 3.1.14.

Assume first that the Boolean function of example 3.1.14 is given by its truth vector, i.e. :

$$[\ f_{\underline{e}}] = [0\ 0\ 1\ 1\ 0\ 1\ 0\ 1]$$

We deduce successively :

$$\underline{\psi}^{(\uparrow)}_{x_2}(f) = \begin{bmatrix} 0 & 0 & 1 & 1 & \underline{0} & \emptyset & \emptyset & \underline{1} \\ \underline{0} & \emptyset & \emptyset & \underline{1} & 0 & 1 & 0 & 1 \end{bmatrix}$$

$$\underline{\psi}^{(\uparrow)}_{x_2x_1}(f) = \begin{bmatrix} 0 & 0 & \emptyset & \emptyset & \underline{0} & \emptyset & \emptyset & \emptyset \\ \emptyset & \emptyset & 1 & 1 & \emptyset & \emptyset & \emptyset & \underline{1} \\ \underline{0} & \emptyset & \emptyset & \emptyset & 0 & 1 & \underline{0} & \underline{1} \\ \emptyset & \emptyset & \emptyset & \underline{1} & \underline{0} & \underline{1} & 0 & 1 \end{bmatrix}$$

$$\underline{\psi}^{(\uparrow)}(f) = \begin{bmatrix} 0 & \underline{0} & \emptyset & e & \underline{0} & \emptyset & \emptyset & e \\ \underline{0} & 0 & e & \emptyset & \emptyset & \emptyset & e & \emptyset \\ \emptyset & e & 1 & \underline{1} & \emptyset & e & \emptyset & \emptyset \\ e & \emptyset & \underline{1} & 1 & e & \emptyset & \emptyset & \underline{1} \\ \underline{0} & \emptyset & \emptyset & e & 0 & \emptyset & \underline{0} & e \\ \emptyset & \emptyset & e & \emptyset & \emptyset & 1 & e & \underline{1} \\ \emptyset & e & \emptyset & \emptyset & \underline{0} & e & 0 & \emptyset \\ e & \emptyset & \emptyset & \underline{1} & e & \underline{1} & \emptyset & 1 \end{bmatrix}$$

From $\underline{\psi}^{(\uparrow)}(f)$ we deduce e.g. :

$$\underline{\psi}^{(\wedge)}(f) = \underline{\psi}^{(\uparrow)}_{(\emptyset=e=0)}(f) = \begin{bmatrix} 0 \\ 0 \\ x_1\bar{x}_2 \\ x_1(x_0 \vee \bar{x}_2) \\ 0 \\ x_0x_2 \\ 0 \\ x_0(x_1 \vee x_2) \end{bmatrix} \quad ; \quad \underline{\psi}^{(\vee)}(f) = \underline{\psi}^{(\uparrow)}_{(\emptyset=e=1)}(f) = \begin{bmatrix} x_1 \vee x_0x_2 \\ x_1 \vee x_2 \\ 1 \\ 1 \\ x_0 \vee x_1\bar{x}_2 \\ 1 \\ x_0 \vee \bar{x}_2 \\ 1 \end{bmatrix}$$

Lower envelopes ; Upper envelopes

$$\underline{\psi}^{(\oplus)}(f) = \underline{\psi}^{(\uparrow)}_{(\emptyset=1,e=\underline{1}=0)} = \begin{bmatrix} 0 & 0 & 1 & 0 & 0 & 1 & 1 & 0 \\ 0 & 0 & 0 & 1 & 1 & 1 & 0 & 1 \\ 1 & 0 & 1 & 0 & 1 & 0 & 1 & 1 \\ 0 & 1 & 0 & 1 & 0 & 1 & 1 & 0 \\ 0 & 1 & 1 & 0 & 0 & 1 & 0 & 0 \\ 1 & 1 & 0 & 1 & 1 & 1 & 0 & 0 \\ 1 & 0 & 1 & 1 & 0 & 0 & 0 & 1 \\ 0 & 1 & 1 & 0 & 0 & 0 & 1 & 1 \end{bmatrix} \qquad \begin{matrix} (0\ 0\ 0) \\ (0\ 0\ 1) \\ (0\ 1\ 0) \\ (0\ 1\ 1) \\ (1\ 0\ 0) \\ (1\ 0\ 1) \\ (1\ 1\ 0) \\ (1\ 1\ 1) \end{matrix}$$

$\underline{h}$-transforms ; $\underline{h}$

From the $\underline{h}$-transforms we deduce the Reed-Muller expansions (or Newton expansions of f at $\underline{x}=\underline{h}$, i.e. :

$$\begin{bmatrix} x_0x_2 \oplus x_1 \oplus x_1x_2 \\ x_2 \oplus x_2\bar{x}_0 \oplus x_1 \oplus x_1x_2 \\ 1 \oplus x_2 \oplus x_2x_0 \oplus \bar{x}_1 \oplus \bar{x}_1x_2 \\ 1 \oplus x_2\bar{x}_0 \oplus \bar{x}_1 \oplus \bar{x}_1x_2 \\ x_0 \oplus \bar{x}_2x_0 \oplus \bar{x}_2x_1 \\ 1 \oplus \bar{x}_0 \oplus \bar{x}_0\bar{x}_2 \oplus \bar{x}_2 \oplus \bar{x}_2x_1 \\ \bar{x}_2 \oplus x_0 \oplus x_0\bar{x}_2 \oplus \bar{x}_1\bar{x}_2 \\ 1 \oplus \bar{x}_0 \oplus \bar{x}_0\bar{x}_2 \oplus \bar{x}_1\bar{x}_2 \end{bmatrix} \qquad \begin{matrix} (0\ 0\ 0) \\ (0\ 0\ 1) \\ (0\ 1\ 0) \\ (0\ 1\ 1) \\ (1\ 0\ 0) \\ (1\ 0\ 1) \\ (1\ 1\ 0) \\ (1\ 1\ 1) \end{matrix}$$

Reed-Muller expansions at $\underline{h}$; $\underline{h}$

Observe finally that performing the operation $\underline{\psi}^{(\uparrow)}(f)$ does not request that the function f be given by its truth vector : the computation may start from any formal expression giving f.

For the example we deduce :

$$f = x_0x_2 \vee x_1\bar{x}_2$$

$$\underline{\psi}^{(\uparrow)}_{x_2}(f) = \left[\begin{array}{c|c} x_1 & x_0 \uparrow x_1 \\ \hline x_0 \uparrow x_1 & x_0 \end{array}\right]$$

$$\underline{\psi}^{(\uparrow)}_{x_2x_1}(f) = \left[\begin{array}{c|c|c|c} 0 & \emptyset & x_0 \uparrow 0 & e(x_0)\uparrow\emptyset \\ \hline \emptyset & 1 & e(x_0)\uparrow\emptyset & x_0 \uparrow 1 \\ \hline x_0 \uparrow 0 & e(x_0)\uparrow\emptyset & x_0 & e(x_0) \\ \hline e(x_0)\uparrow\emptyset & x_0 \uparrow 1 & e(x_0) & x_0 \end{array}\right]$$

$$\underline{\psi}^{(\uparrow)}_{x_2x_1x_0}(f) = \underline{\psi}^{(\uparrow)}(f) \; .$$

4.2.2.4. Obtention of the matrix $\underline{\psi}^{(\uparrow)}(f)$ from a diagonal matrix.

In this section we present a somewhat different method for obtaining the $(2^n\times 2^n)$ matrix $\underline{\psi}^{(\uparrow)}(f)$. Let $D_0(f)$ be the $(2^n\times 2^n)$ matrix formed by the truth vector of f on its main diagonal ; the other elements of $D_0(f)$ are not defined. For the example of the preceding section we have :

$$D_0(f) = \begin{bmatrix} 0 & - & - & - & - & - & - & - \\ - & 0 & - & - & - & - & - & - \\ - & - & 1 & - & - & - & - & - \\ - & - & - & 1 & - & - & - & - \\ - & - & - & - & 0 & - & - & - \\ - & - & - & - & - & 1 & - & - \\ - & - & - & - & - & - & 0 & - \\ - & - & - & - & - & - & - & 1 \end{bmatrix}$$

The matrix $D_1^{(\uparrow)}(f)$ if obtained from the matrix $D_0(f)$ as follows : for each index $i \in \{0,2,\ldots,2^{n-1}\}$, representing the row and column indices of $D_0^{(\uparrow)}(f)$, perform the following substitution

$$\begin{array}{c} \\ i \\ i+1 \\ \end{array} \left[\begin{array}{c|c} a_i & \\ \hline & a_{i+1} \end{array}\right]_{\;i \quad i+1} \;\rightarrow\; \left[\begin{array}{c|c} a_i & a_i \uparrow a_{i+1} \\ \hline a_i \uparrow a_{i+1} & a_{i+1} \end{array}\right]_{\;i \quad i+1} \begin{array}{c} i \\ i+1 \end{array}$$

The matrix $D_2^{(\uparrow)}(f)$ is obtained from $D_1^{(\uparrow)}(f)$ by performing the same operation on the (2×2) diagonal matrices of $D_1^{(\uparrow)}(f)$; in general the matrix $D_i^{(\uparrow)}(f)$ is obtained from $D_{i-1}^{(\uparrow)}(f)$ by considering the $(2^{i-1}\times 2^{i-1})$ diagonal matrices of $D_{i-1}^{(\uparrow)}(f)$.

<u>Theorem</u>

$$D_n^{(\uparrow)}(f) = \underline{\psi}^{(\uparrow)}(f)$$

<u>Proof</u>

The proof derives from the fact that the induction step for obtaining $D_n^{(\uparrow)}(f)$ is nothing but (4.14) which was the induction step for obtaining $\underline{\psi}^{(\uparrow)}(f)$. □

Consider again the illustrative example of section 4.2.2.3; we obtain successively :

$$D_0(f) = \begin{bmatrix} 0 &&&&&&& \\ & 0 &&&&&& \\ && 1 &&&&& \\ &&& 1 &&&& \\ &&&& 0 &&& \\ &&&&& 1 && \\ &&&&&& 0 & \\ &&&&&&& 1 \end{bmatrix}
\qquad
D_1^{(\uparrow)}(f) = \begin{bmatrix} 0 & \underline{0} &&&&&& \\ \underline{0} & 0 &&&&&& \\ && 1 & \underline{1} &&&& \\ && \underline{1} & 1 &&&& \\ &&&& 0 & \emptyset && \\ &&&& \emptyset & 1 && \\ &&&&&& 0 & \emptyset \\ &&&&&& \emptyset & 1 \end{bmatrix}$$

$$D_2^{(\uparrow)}(f) = \begin{bmatrix} 0 & \underline{0} & \emptyset & e &&&& \\ \underline{0} & 0 & e & \emptyset &&&& \\ \emptyset & e & 1 & \underline{1} &&&& \\ e & \emptyset & \underline{1} & 1 &&&& \\ &&&& 0 & \emptyset & \underline{0} & e \\ &&&& \emptyset & 1 & e & \underline{1} \\ &&&& \underline{0} & e & 0 & \emptyset \\ &&&& e & \underline{1} & \emptyset & 1 \end{bmatrix}
\qquad
D_3^{(\uparrow)}(f) = \underline{\psi}^{(\uparrow)}(f) = \begin{bmatrix} 0 & \underline{0} & \emptyset & e & \underline{0} & \emptyset & \emptyset & e \\ \underline{0} & 0 & e & \emptyset & \emptyset & \emptyset & e & \emptyset \\ \emptyset & e & 1 & \underline{1} & \emptyset & e & \emptyset & \emptyset \\ e & \emptyset & \underline{1} & 1 & e & \emptyset & \emptyset & \underline{1} \\ \underline{0} & \emptyset & \emptyset & e & 0 & \emptyset & \underline{0} & e \\ \emptyset & \emptyset & e & \emptyset & \emptyset & 1 & e & \underline{1} \\ \emptyset & e & \emptyset & \emptyset & \underline{0} & e & 0 & \emptyset \\ e & \emptyset & \emptyset & \underline{1} & e & \underline{1} & \emptyset & 1 \end{bmatrix}$$

4.3. <u>Particular algorithms related to circuit analysis synthesis methods</u>

4.3.1. <u>Synthesis of two-level circuits using AND- and OR-gates</u>

4.3.1.1. <u>Problem statement</u>

The class of two-level networks is perhaps the most important of all. The main objective of switching design is to produce a circuit having the required switching characteristics but utilizing the minimum number of components , either in gates, or in gate inputs.

The synthesis of minimal two-level networks is a classical problem ; we consider two levels, the first level being formed by AND-gates (resp. OR-gates) and the second level being formed by OR-gates (resp. AND-gates). We are interested in the optimal realization of a Boolean function, i.e. a realization containing a minimal number of gates. The solution to this problem may be found in any book dealing with switching theory and logical design (see e.g. Mc Cluskey [1965]) : a circuit will be optimal if each of its AND-gates (resp. OR-gates) realizes a prime implicant (resp. prime implicate) of an irredundant cover for the function f. Stated otherwise, in any optimal synthesis the AND-gates (resp. OR-gates) are in one-to-one correspondence with the prime implicants (resp. prime implicates) of the irredundant cover.

Let us also point out another type of (non-optimal) synthesis directly connected to the preceding one : the AND-gates (resp. OR-gates) are in one-to-one correspondence with *all* prime implicants (resp. prime implicates) of the function f. The importance of this synthesis has been made obvious by Huffman [1957] : the networks built up according to this scheme present a minimum number of logic static hazards (see also Davio, Deschamps and Thayse [1978]).

The two above types of synthesis make obvious the importance of obtaining algorithms allowing us to detect either all prime implicants, or all prime implicates of a Boolean function. In the following sections some of these algorithms which can be derived by making use of the functional properties of the meet differences are studied. We shall only consider the obtention of prime implicants, the obtention of prime implicates being a dual problem which can be stated in a similar way.

4.3.1.2. Algorithm

Assume that the function f is given as a disjunction of cubes (*normal disjunctive form*).

Step 1

- From the expression of f deduce the expressions giving the :

$$\frac{pf}{p\underline{x}^{\underline{k}}},\ \underline{x}=(x_{n-1}, \ldots, x_1, x_0),\ \underline{k}=(k_{n-1}, \ldots, k_1, k_0),\ k_i \in \{0,1\}\ \forall i \quad ,$$

for each $\underline{k}$ of weight $w(\underline{k})=1$ (where the weight counts the coordinates equal to 1). These expressions are obtained from the definition of the pf/px_i, i.e. :

$$pf/px_i = f(x_i=0) \wedge f(x_i=1).$$

- Using the distributivity law, obtain each these $pf/p\underline{x}^{\underline{k}}$ as a disjunction of cubes.

Step j

- From the expressions of $pf/p\underline{x}^{\underline{k}}$, ($\underline{k}$: $w(\underline{k})=j-1$), deduce the expressions of :

$$\frac{pf}{p\underline{x}^{\underline{k}}} , \quad \underline{k} : w(\underline{k})=j \quad .$$

These expressions are obtained from the recurrence (3.45).

- Using distributivity, obtain each of the $pf/p\underline{x}^{\underline{k}}$ as a disjunction of cubes.

Step n

- Obtain $pf/p\underline{x}$ from any one of the expressions $pf/p\underline{x}^{\underline{k}}$ ($w(\underline{k})=n-1$), and by application of the distributivity law, obtain this last expression as a disjunction of cubes.

Step n+1

- The prime implicants of f are obtained by application of the absorption law from the cubes (which are the implicants of the function f) resulting from the n first steps.

4.3.1.3. Theorem

Let $\underline{x}_1, \underline{x}_0$ *be a partition of* $\underline{x}$;

(a) *The minterms of the function :*

$$p_{\underline{x}_0}(f) = \frac{pf}{p\underline{x}_0} \bigwedge_{x_i \in \underline{x}_1} \overline{\frac{pf}{p\underline{x}_0 x_i}} \tag{4.21}$$

are the prime implicants of f *degenerate in* $\underline{x}_0$ *and containing (under either direct or complemented form) each of the variables* $x_i \in \underline{x}_1$.

(b) *The minterms of the function :*

$$q_{\underline{x}_0}(f) = \overline{\frac{qf}{q\underline{x}_0}} \bigwedge_{x_i \in \underline{x}_0} \frac{qf}{q\underline{x}_0 x_i} \tag{4.22}$$

are the prime implicants of $\bar{f}$ *degenerate in* $\underline{x}_0$ *and containing (either under direct or under complemented form) each of the variables* $x_i \in \underline{x}_1$.

Proof

(a) Since $pf/p\underline{x}_0$ is the disjunction of the prime implicants of f degenerate in $\underline{x}_0$, if we multiply this function by $\bigwedge_i \overline{pf/p\underline{x}_0 x_i}$, we eliminate the prime implicants which do not depend effectively on all variables in $\underline{x}_1$.

(b) Dual statement. □

From this theorem we deduce the following algorithm for computing all prime implicants of f.

4.3.1.4. Algorithm

Step 1.

- Starting from any well-formed expression of f, compute all the $pf/p\underline{x}^{\underline{k}}$, $k_i \in \{0,1\}$ $\forall i$.

These meet differences are obtained e.g. by means of the recurrence formula (3.45) by considering the successive vectors $\underline{k}$ ordered with increasing weights. Otherwise stated, we start from $pf/p\underline{x}^{\underline{0}} = f$ $(w(\underline{k})=0)$ for reaching $pf/p\underline{x}(w(\underline{k})=n)$. Each of these $pf/p\underline{x}^{\underline{k}}$ is obtained as a well-formed expression .

Step 2.

- The prime implicants are obtained from the $pf/p\underline{x}^{\underline{k}}$ computed at the step 1 and from theorem 4.3.1.3.

Comments

Let us compare the algorithms 4.3.1.2. and 4.3.1.4.

- In algorithm 4.3.1.2. the initial forms of the function and of each of the $pf/p\underline{x}^{\underline{k}}$ are the normal disjunctive forms. In algorithm 4.3.1.4, the function f and the $pf/p\underline{x}^{\underline{k}}$ are given by the intermediate of a well-formed expression.
- In algorithm 4.3.1.2. the prime implicants are obtained from the implicants by application of the absorption rule (step n+1). The final result of algorithm 4.3.1.4. is formed by the prime implicants only. The comparison of numerous implicants of f for the application of the absorption rule is thus avoided. The computation resulting from the absorption rule is replaced by the computation of the functions $p_{\underline{x}_0}(f)$ (step 2 in 4.3.1.4.).

4.3.1.5. Computation method for obtaining $pf/p\underline{x}^{\underline{k}}$ and $qf/q\underline{x}^{\underline{k}}$

The algorithms 4.3.1.2. and 4.3.1.4. are grounded on computation methods for the $pf/p\underline{x}^{\underline{k}}$ deriving from the definition of these functions. We suggest in this section a completely different approach for the computation of these functions. Let us define the operator rf/rx_i in the following way :

$$\frac{rf}{rx_i} = f(x_i=f) \ . \tag{4.23}$$

Expanding (4.23) with respect to the variable x_i, we successively obtain :

$$\begin{aligned}\frac{rf}{rx_i} &= \bar{f}\ f(x_i=0) \vee f\ f(x_i=1) \\ &= \frac{pf}{px_i} \vee x_i \frac{qf}{qx_i} \ .\end{aligned} \tag{4.24}$$

We define the multiple operator $rf/r\underline{x}_0$ as follows :

$$\frac{rf}{r\underline{x}_0} = \frac{r}{rx_{p-1}} (\ldots \frac{r}{rx_1} (\frac{rf}{rx_0}) \ldots), \ \underline{x}_0 = (x_{p-1}, \ldots, x_1, x_0) \quad . \tag{4.25}$$

From (4.24) and (4.25) we deduce that $rf/r\underline{x}_0$ may be expressed in the following symbolic form :

$$\frac{rf}{r\underline{x}_0} = \bigwedge_{i=p-1,0} (\frac{p}{px_i} \vee x_i \frac{q}{qx_i}) \ f \ . \tag{4.26}$$

We obtain finally the following formulas :

$$\frac{pf}{p\underline{x}_0} = (\frac{rf}{r\underline{x}_0})_{\underline{x}_0 = \underline{0}} \ ,$$

$$\frac{qf}{q\underline{x}_0} = (\frac{rf}{r\underline{x}_0})_{\underline{x}_0 = \underline{1}} \ . \tag{4.27}$$

If we observe that :

$$\frac{r}{rx_i} (\frac{rf}{rx_j}) \neq \frac{r}{rx_j} (\frac{rf}{rx_i}) \ ,$$

while :

$$(\frac{r}{rx_i} (\frac{rf}{rx_j}))_{(x_i,x_j)=(e,e)} = (\frac{r}{rx_j} (\frac{rf}{rx_i}))_{(x_i,x_j)=(e,e)} \ , \ e \in \{0,1\}$$

it appears that the ordering of the successive "derivations" in $rf/r\underline{x}_0$ is unimportant if and only if the expected result is the computation of $pf/p\underline{x}_0$ and of $qf/q\underline{x}_0$.

4.3.1.6. Algorithm.

We present in this section an algorithm which is based on the use of the simple meet differences pf/px_i ; the main interest of this algorithm lies in the fact that it does not request the computation of higher order differences of the form $pf/p\underline{x}_0$. This algorithm may be considered as the description of the generalized consensus algorithm by means of the concept of meet difference.

First of all, let us notethat, since $pf/px_i \leqslant f$ holds, one has :

$$f \vee \frac{pf}{px_i} = f.$$

Let us now assume that f be given as a disjunction of cubes ; these cubes may form a completely random covering of the function. If pf/px_i is also given as a disjunctio of cubes, we know that the expression $f \vee pf/px_i$ is equal to f and contains certainly all cubes of f degenerate in the variable x_i. Thus, while functionally equivalent, the expressions f and $f \vee pf/px_i$ are not identical since they may contain different sets of cubes. Further on in this section, we shall assume that an expression such as $f \vee pf/px_i$ not only represents an expression functionally equivalent to f but has also the meaning, quoted above, of a disjunction of cubes. The following relation holds :

$$p(f \vee pf/px_i)/px_j = [f(x_j=0) \vee pf(x_j=0)/px_i] \wedge [f(x_j=1) \vee pf(x_j=1)/px_i)]$$

$$= pf/px_i \vee pf/px_i x_j \quad .$$

More generally, one has for a function $f(\underline{x},y)$:

$$p(f \vee \bigvee_{\underline{e}} pf/p\underline{x}^{\underline{e}})/py = pf/py \vee \bigvee_{\underline{e}} pf/p\underline{x}^{\underline{e}}y \ .$$

On the basis of this observation, the following algorithm may be stated for detecting all prime implicants of the function f.

Step 0 :

Set $f=f_0$.

Step j $(1 \leqslant j \leqslant n)$

From the expression of f_{j-1}, deduce pf_{j-1}/px_{j-1} by application of the formula :

$$pf_{j-1}/px_{j-1} = f_{j-1}\ (x_{j-1}=0) \wedge f_{j-1}(x_{j-1}=1)$$

and obtain pf_{j-1}/px_{j-1} as a disjunction of cubes by applying the distributivity law. The function f_j is defined as the disjunction of the cubes of f_{j-1} and of pf_{j-1}/px_{j-1} in which the cubes contained in other cubes have been deleted by application of the absorption law ; we write symbolically :

$$f_j = f_{j-1} \vee pf_{j-1}/px_{j-1} \quad .$$

The function f_n obtained at the end of the n-th step is the function f written as a disjunction of all its prime implicants.

4.3.1.7. Example

Consider the six-variable switching function f given by the truth table 4.IV and by the disjunction of cubes f_0 :

$$f = f_0 = x_5x_1\bar{x}_0 \vee \bar{x}_4\bar{x}_1\bar{x}_0 \vee x_5x_4\bar{x}_3\bar{x}_1 \vee \bar{x}_5\bar{x}_4x_1\bar{x}_0 \vee \bar{x}_4\bar{x}_3\bar{x}_1x_0 \vee x_5x_3\bar{x}_2\bar{x}_1x_0 \vee$$
$$x_5x_3x_2x_1x_0 \vee \bar{x}_5\bar{x}_4x_3x_2x_1x_0 \vee \bar{x}_5\bar{x}_4x_3\bar{x}_2\bar{x}_1x_0 \vee \bar{x}_5x_4x_3x_2\bar{x}_1x_0 \vee$$
$$\bar{x}_5x_4x_3\bar{x}_2x_1x_0 \vee \bar{x}_5x_4\bar{x}_3x_1x_0 \vee x_5x_4x_3\bar{x}_1\bar{x}_0 .$$

$$pf_0/px_0 = x_5x_4\bar{x}_3\bar{x}_1 \vee x_5x_3x_2x_1 \vee \bar{x}_4\bar{x}_3\bar{x}_1 \vee \bar{x}_5\bar{x}_4x_3\bar{x}_2\bar{x}_1 \vee \bar{x}_5\bar{x}_4x_3x_2x_1 \vee$$
$$x_5x_4x_3\bar{x}_2\bar{x}_1 \vee x_5\bar{x}_4x_3\bar{x}_2\bar{x}_1 .$$

$$f_0 \vee pf_0/px_0 = x_5x_1\bar{x}_0 \vee \bar{x}_4\bar{x}_1\bar{x}_0 \vee x_5x_4\bar{x}_3\bar{x}_1 \vee \bar{x}_5\bar{x}_4x_1\bar{x}_0 \vee x_5x_3\bar{x}_2\bar{x}_1x_0 \vee$$
$$\bar{x}_5x_4x_3x_2\bar{x}_1x_0 \vee \bar{x}_5x_4x_3\bar{x}_2x_1x_0 \vee \bar{x}_5x_4\bar{x}_3x_1x_0 \vee x_5x_4x_3\bar{x}_1\bar{x}_0 \vee$$
$$x_5x_4\bar{x}_3\bar{x}_1 \vee x_5x_3x_2x_1 \vee \bar{x}_4\bar{x}_3\bar{x}_1 \vee \bar{x}_5\bar{x}_4x_3\bar{x}_2\bar{x}_1 \vee \bar{x}_5\bar{x}_4x_3x_2x_1 \vee$$
$$x_5x_4x_3\bar{x}_2\bar{x}_1 \vee x_5\bar{x}_4x_3\bar{x}_2\bar{x}_1 = f_1 .$$

$$pf_1/px_1 = x_5\bar{x}_4\bar{x}_0 \vee \bar{x}_5\bar{x}_4\bar{x}_0 \vee x_5x_4\bar{x}_3\bar{x}_0 \vee x_5x_4x_3\bar{x}_0 .$$

$$f_1 \vee pf_1/px_1 = x_5x_1\bar{x}_0 \vee \bar{x}_4\bar{x}_1\bar{x}_0 \vee x_5x_4\bar{x}_3\bar{x}_1 \vee x_5x_3\bar{x}_2\bar{x}_1x_0 \vee \bar{x}_5x_4x_3x_2\bar{x}_1x_0 \vee$$
$$\bar{x}_5x_4x_3\bar{x}_2x_1x_0 \vee \bar{x}_5x_4\bar{x}_3x_1x_0 \vee x_5x_4\bar{x}_3\bar{x}_1 \vee x_5x_3x_2x_1 \vee \bar{x}_4\bar{x}_3\bar{x}_1 \vee$$
$$\bar{x}_5\bar{x}_4x_3\bar{x}_2\bar{x}_1 \vee \bar{x}_5\bar{x}_4x_3x_2x_1 \vee x_5x_4x_3\bar{x}_2\bar{x}_1 \vee x_5\bar{x}_4x_3\bar{x}_2\bar{x}_1 \vee x_5\bar{x}_4\bar{x}_0 \vee$$
$$\bar{x}_5\bar{x}_4\bar{x}_0 \vee x_5x_4\bar{x}_3\bar{x}_0 \vee x_5x_4x_3\bar{x}_0 = f_2$$

$$pf_2/px_2 = x_5x_1\bar{x}_0 \vee \bar{x}_4\bar{x}_1\bar{x}_0 \vee x_5x_4\bar{x}_3\bar{x}_1 \vee \bar{x}_5x_4\bar{x}_3x_1x_0 \vee x_5x_4\bar{x}_3\bar{x}_1 \vee \bar{x}_4\bar{x}_3\bar{x}_1$$
$$\vee x_5\bar{x}_4\bar{x}_0 \vee \bar{x}_5\bar{x}_4\bar{x}_0 \vee x_5x_4\bar{x}_3\bar{x}_0 \vee x_5x_4x_3\bar{x}_0 .$$

$$f_2 \vee pf_2/px_2 = f_2 = f_3 .$$

$$pf_3/px_3 = x_5x_1\bar{x}_0 \vee \bar{x}_4\bar{x}_1\bar{x}_0 \vee x_5\bar{x}_4\bar{x}_0 \vee \bar{x}_5\bar{x}_4\bar{x}_0 \vee x_5x_4\bar{x}_0 \vee x_5x_4\bar{x}_2\bar{x}_1 \vee$$
$$\bar{x}_5x_4\bar{x}_2x_1x_0 \vee \bar{x}_5\bar{x}_4\bar{x}_2\bar{x}_1 \vee x_5\bar{x}_4\bar{x}_2\bar{x}_1 .$$

$$f_3 \vee pf_3/px_3 = x_5x_1\bar{x}_0 \vee \bar{x}_4\bar{x}_1\bar{x}_0 \vee x_5\bar{x}_4\bar{x}_0 \vee \bar{x}_5\bar{x}_4\bar{x}_0 \vee x_5x_4\bar{x}_0 \vee x_5x_4\bar{x}_2\bar{x}_1 \vee$$
$$\bar{x}_5x_4\bar{x}_2x_1x_0 \vee \bar{x}_5\bar{x}_4\bar{x}_2\bar{x}_1 \vee x_5x_3\bar{x}_2\bar{x}_1x_0 \vee \bar{x}_5x_4x_3x_2\bar{x}_1x_0 \vee$$
$$\bar{x}_5x_4\bar{x}_3x_1x_0 \vee x_5x_4\bar{x}_3\bar{x}_1 \vee x_5x_3x_2x_1 \vee \bar{x}_4\bar{x}_3\bar{x}_1 \vee \bar{x}_5\bar{x}_4x_3x_2x_1 = f_4 .$$

$$pf_4/px_4 = x_5\bar{x}_0 \vee x_5\bar{x}_2\bar{x}_1 \vee x_5\bar{x}_3\bar{x}_1$$

$$\begin{aligned}
f_4 \vee pf_4/px_4 \;=\;& x_5\bar{x}_0 \vee x_5\bar{x}_2\bar{x}_1 \vee x_5\bar{x}_3\bar{x}_1 \vee x_5x_3x_2x_1 \vee \bar{x}_4\bar{x}_1\bar{x}_0 \vee \bar{x}_5\bar{x}_4\bar{x}_0 \vee \\
& \bar{x}_5x_4\bar{x}_2x_1x_0 \vee \bar{x}_5\bar{x}_4\bar{x}_2\bar{x}_1 \vee \bar{x}_5x_4x_3x_2\bar{x}_1x_0 \vee \bar{x}_5x_4\bar{x}_3x_1x_0 \vee \\
& \bar{x}_4\bar{x}_3\bar{x}_1 \vee \bar{x}_5\bar{x}_4x_3x_2x_1 = f_5 .
\end{aligned}$$

$$pf_5/px_5 \;=\; \bar{x}_4\bar{x}_0 \vee \bar{x}_4\bar{x}_3\bar{x}_1 \vee \bar{x}_4\bar{x}_2\bar{x}_1 \vee \bar{x}_4x_3x_2x_1 .$$

$$\begin{aligned}
f_5 \vee pf_5/px_5 \;=\;& x_5\bar{x}_0 \vee \bar{x}_4\bar{x}_0 \vee \bar{x}_4\bar{x}_3\bar{x}_1 \vee \bar{x}_4\bar{x}_2\bar{x}_1 \vee x_5\bar{x}_2\bar{x}_1 \vee x_5\bar{x}_3\bar{x}_1 \vee \\
& x_5x_3x_2x_1 \vee \bar{x}_4x_3x_2x_1 \vee \bar{x}_5x_4\bar{x}_2x_1x_0 \vee \bar{x}_5x_4\bar{x}_3x_1x_0 \vee \bar{x}_5x_4x_3x_2\bar{x}_1x_0 , \\
=\;& f_6 .
\end{aligned}$$

$x_4x_5 = 0\ 0$ (x_0 above the middle columns; x_2 beside the middle rows)

1	1	0	1
1	1	0	1
1	0	1	1
1	1	0	1

$x_4x_5 = 0\ 1$ (x_0 above the middle columns; x_3 beside the lower rows)

1	1	0	1
1	1	0	1
1	0	1	1
1	1	0	1

$x_4x_5 = 1\ 0$ (x_2 beside the middle rows; x_1 below the right columns)

0	0	1	0
0	0	1	0
0	1	0	0
0	0	1	0

$x_4x_5 = 1\ 1$ (x_3 beside the lower rows; x_1 below the right columns)

1	1	0	1
1	1	0	1
1	0	1	1
1	1	0	1

Table 4.IV.

Let us illustrate the algorithm 4.3.1.4. and the various computation methods for the functions $pf/p\underline{x}_0$. We consider again the same function f but expressed now as a modulo-2 sum of cubes :

$$f = 1 \oplus x_1x_0 \oplus x_4 \oplus x_5x_4 \oplus x_3x_2x_0 . \tag{4.28}$$

From this expression we compute the functions $rf/r\underline{x}_0$; these functions are gathered in Table 4.V. For the sake of conciseness, we have replaced the letters x_i and $\bar{x}_i$, in the entries of table 4.V. by the symbols i and $\bar{\text{i}}$ respectively ; consequently we have also replaced the symbols 0 and 1 by ∅ and I respectively. The functions $pf/p\underline{x}_0$ and $qf/q\underline{x}_0$, which may either be deduced from $rf/r\underline{x}_0$ (formulas (4.27)) or be obtained from their definition and from (4.8), have been gathered in Table 4.VI.

$\underline{x}_0$	$rf/r\underline{x}_0$	$\underline{x}_0$	$rf/r\underline{x}_0$
∅	$I\oplus10\oplus\bar{5}4\oplus320$	2 3	$I\oplus\bar{5}4\oplus\bar{5}430\oplus\bar{5}4\bar{2}0\oplus3\bar{2}\bar{1}0\oplus\bar{3}210\oplus\bar{2}0$
0	$I\oplus1\bar{0}\oplus\bar{5}4\bar{1}\oplus32\bar{0}\oplus\bar{5}432$	2 4	$I\oplus5\bar{3}10\oplus3\bar{2}0\oplus\bar{5}\bar{4}\oplus\bar{5}\bar{4}30$
1	$I\oplus\bar{1}0\oplus\bar{5}4\bar{0}$	2 5	$I\oplus3\bar{2}0\oplus\bar{5}4\oplus\bar{5}430\oplus\bar{4}\bar{3}10$
2	$I\oplus\bar{3}10\oplus\bar{5}4\oplus\bar{5}430\oplus3\bar{2}0$	3 4	$I\oplus5\bar{2}10\oplus\bar{3}20\oplus\bar{5}\bar{4}\oplus\bar{5}\bar{4}20$
3	$I\oplus\bar{2}10\oplus\bar{5}4\oplus\bar{5}420\oplus\bar{3}20$	3 5	$I\oplus\bar{3}20\oplus\bar{4}\bar{2}10\oplus\bar{5}4\oplus\bar{5}420$
4	$5\oplus510\oplus5320\oplus\bar{5}4$	4 5	$5\oplus\bar{5}4\oplus510\oplus410\oplus4320\oplus5320$
5	$I\oplus\bar{4}10\oplus\bar{5}4\oplus\bar{4}320$	0 1 2	$I\oplus\bar{1}\bar{0}\oplus\bar{5}4\bar{1}0\oplus\bar{5}41\bar{0}$
0 1	$I\oplus\bar{1}\bar{0}\oplus\bar{5}4\bar{1}0\oplus\bar{5}41\bar{0}$	0 1 3	$I\oplus\bar{1}\bar{0}\oplus\bar{5}4\bar{1}0\oplus\bar{5}41\bar{0}$
0 2	$I\oplus\bar{5}4\oplus\bar{3}1\bar{0}\oplus\bar{5}4\bar{3}1\oplus\bar{5}430\oplus\bar{5}43\bar{2}\oplus3\bar{2}\bar{0}$	0 1 4	$I\oplus\bar{1}\bar{0}\oplus\bar{5}\bar{4}\bar{1}0\oplus\bar{5}\bar{4}1\bar{0}$
0 3	$I\oplus\bar{5}4\oplus\bar{2}1\bar{0}\oplus\bar{5}4\bar{2}1\oplus\bar{5}420\oplus\bar{5}4\bar{3}2\oplus\bar{3}2\bar{0}$	0 1 5	$I\oplus\bar{1}\bar{0}\oplus\bar{5}4\bar{1}0\oplus\bar{5}41\bar{0}$
0 4	$I\oplus1\bar{0}\oplus32\bar{0}\oplus\bar{5}\bar{4}\bar{1}\oplus\bar{5}\bar{4}32$	0 2 3	$I\oplus\bar{3}1\bar{0}\oplus3\bar{2}\bar{0}\oplus\bar{2}\bar{1}\bar{0}\oplus\bar{5}4(I\oplus\bar{3}1\oplus320\oplus\bar{3}\bar{2}\bar{0}\oplus\bar{2}1)$
0 5	$I\oplus1\bar{0}\oplus32\bar{0}\oplus\bar{5}4\bar{1}\oplus\bar{5}432$	0 2 4	$I\oplus\bar{3}1\bar{0}\oplus3\bar{2}\bar{0}\oplus\bar{5}\bar{4}\oplus\bar{5}\bar{4}30\oplus\bar{5}\bar{4}\bar{3}1\oplus\bar{5}\bar{4}3\bar{2}$
1 2	$I\oplus\bar{1}0\oplus\bar{5}4\bar{0}$	0 2 5	$I\oplus\bar{3}1\bar{0}\oplus3\bar{2}\bar{0}\oplus\bar{5}4\oplus\bar{5}4\bar{3}1\oplus\bar{5}43\bar{2}\oplus\bar{5}430$
1 3	$I\oplus\bar{1}0\oplus\bar{5}4\bar{0}$	0 3 4	$I\oplus\bar{2}1\bar{0}\oplus\bar{3}2\bar{0}\oplus\bar{5}\bar{4}\oplus\bar{5}\bar{4}20\oplus\bar{5}\bar{4}\bar{2}1\oplus\bar{5}\bar{4}\bar{3}2$
1 4	$I\oplus\bar{1}0\oplus\bar{5}\bar{4}\bar{0}$	0 3 5	$I\oplus\bar{2}1\bar{0}\oplus\bar{3}2\bar{0}\oplus\bar{5}4\oplus\bar{5}4\bar{2}1\oplus\bar{5}4\bar{3}2\oplus\bar{5}420$
1 5	$I\oplus\bar{1}0\oplus\bar{5}4\bar{0}$	0 4 5	$I\oplus1\bar{0}\oplus32\bar{0}\oplus\bar{5}\bar{4}\bar{1}\oplus\bar{5}\bar{4}32$

Table 4.V.

$\underline{x}_0$	$rf/r\underline{x}_0$	$\underline{x}_0$	$rf/r\underline{x}_0$
1 2 3	$\mathrm{I}\oplus\bar{1}0\oplus\bar{5}4\bar{0}$	0 2 3 4	$\mathrm{I}\oplus\bar{3}1\bar{0}\oplus53\bar{2}\bar{0}\oplus5\bar{2}\bar{1}\bar{0}\oplus\bar{5}\bar{4}(\mathrm{I}\oplus\bar{3}1\oplus320)\oplus\bar{5}4\bar{2}(\bar{3}\bar{0}\oplus1\oplus\bar{3}10)$
1 2 4	$\mathrm{I}\oplus\bar{1}0\oplus\bar{5}\bar{4}\bar{0}$	0 2 3 5	$\mathrm{I}\oplus\bar{3}1\bar{0}\oplus\bar{4}\bar{2}\bar{1}\bar{0}\oplus4\bar{3}\bar{2}1\oplus4\bar{5}(1\oplus320\oplus\bar{2}1\oplus\bar{3}21\oplus\bar{3}\bar{2}\bar{1}\bar{0})\oplus 3\bar{2}\bar{0}$
1 2 5	$\mathrm{I}\oplus\bar{1}0\oplus\bar{5}4\bar{0}$	0 2 4 5	$\mathrm{I}\oplus\bar{3}1\bar{0}\oplus3\bar{2}\bar{0}\oplus\bar{5}\bar{4}(\mathrm{I}\oplus30\oplus\bar{3}1\oplus3\bar{2})$
1 3 4	$\mathrm{I}\oplus\bar{1}0\oplus\bar{5}\bar{4}\bar{0}$	0 3 4 5	$\mathrm{I}\oplus\bar{5}\bar{4}\oplus\bar{3}2\bar{0}\oplus\bar{5}\bar{4}\bar{3}2\oplus\bar{5}\bar{2}1\bar{0}\oplus\bar{5}\bar{4}\bar{2}1\oplus\bar{5}\bar{4}20$
1 3 5	$\mathrm{I}\oplus\bar{1}0\oplus\bar{5}4\bar{0}$	1 2 3 4	$\mathrm{I}\oplus\bar{1}0\oplus\bar{5}\bar{4}\bar{0}$
1 4 5	$\mathrm{I}\oplus\bar{1}0\oplus\bar{5}\bar{4}\bar{0}$	1 2 3 5	$\mathrm{I}\oplus\bar{1}0\oplus\bar{5}4\bar{0}$
2 3 4	$\mathrm{I}\oplus\bar{5}\bar{4}(\mathrm{I}\oplus30\oplus\bar{2}0)\oplus\bar{2}0(\mathrm{I}\oplus\bar{5}3\oplus53\bar{1})\oplus5\bar{3}210$	1 2 4 5	$\mathrm{I}\oplus\bar{1}0\oplus\bar{5}\bar{4}\bar{0}$
2 3 5	$\mathrm{I}\oplus\bar{5}4(\mathrm{I}\oplus30\oplus\bar{2}0)\oplus\bar{2}0(\mathrm{I}\oplus43\oplus\bar{4}3\bar{1})\oplus\bar{4}\bar{3}210$	1 3 4 5	$\mathrm{I}\oplus\bar{3}20\oplus\bar{5}\bar{4}\bar{0}\oplus\bar{5}\bar{2}\bar{1}0$
2 4 5	$\mathrm{I}\oplus3\bar{2}0\oplus\bar{5}\bar{3}10\oplus\bar{4}\bar{3}10\oplus\bar{5}\bar{4}\oplus\bar{5}\bar{4}30$	3 4 5 2	$\mathrm{I}\oplus\bar{5}\bar{4}\oplus\bar{5}\bar{4}\bar{3}0\oplus\bar{3}\bar{2}0\oplus(\bar{4}\oplus\bar{5})(\bar{2}\oplus\bar{3})10$
3 4 5	$\mathrm{I}\oplus\bar{3}20\oplus\bar{5}\bar{2}10\oplus\bar{5}\bar{4}\oplus\bar{5}\bar{4}20\oplus\bar{4}\bar{2}10$	0 1 2 3 4	$\mathrm{I}\oplus\bar{1}\bar{0}\oplus\bar{5}\bar{4}\bar{1}0\oplus\bar{5}\bar{4}1\bar{0}$
0 1 2 3	$\mathrm{I}\oplus\bar{1}\bar{0}\oplus\bar{5}4\bar{1}0\oplus\bar{5}41\bar{0}$	0 1 2 3 5	$\mathrm{I}\oplus\bar{1}\bar{0}\oplus\bar{5}4\bar{1}0\oplus\bar{5}41\bar{0}$
0 1 2 4	$\mathrm{I}\oplus\bar{1}\bar{0}\oplus\bar{5}\bar{4}\bar{1}0\oplus\bar{5}\bar{4}1\bar{0}$	0 1 2 4 5	$\mathrm{I}\oplus\bar{1}\bar{0}\oplus\bar{5}\bar{4}\bar{1}0\oplus\bar{5}\bar{4}1\bar{0}$
0 1 2 5	$\mathrm{I}\oplus\bar{1}\bar{0}\oplus\bar{5}4\bar{1}0\oplus\bar{5}41\bar{0}$	0 1 3 4 5	$\mathrm{I}\oplus\bar{1}\bar{0}\oplus\bar{5}\bar{4}\bar{1}0\oplus\bar{5}\bar{4}1\bar{0}$
0 1 3 4	$\mathrm{I}\oplus\bar{1}\bar{0}\oplus\bar{5}\bar{4}\bar{1}0\oplus\bar{5}\bar{4}1\bar{0}$	0 2 3 4 5	$\mathrm{I}\oplus\bar{5}\bar{4}(\mathrm{I}\oplus\bar{3}0\oplus\bar{3}\bar{2})\oplus(\bar{5}\oplus\bar{4})(\bar{3}\oplus\bar{2})1\bar{0}\oplus\bar{3}\bar{2}\bar{0}$
0 1 3 5	$\mathrm{I}\oplus\bar{1}\bar{0}\oplus\bar{5}4\bar{1}0\oplus\bar{5}41\bar{0}$	1 2 3 4 5	$\mathrm{I}\oplus\bar{1}0\oplus\bar{5}\bar{4}\bar{0}$
0 1 4 5	$\mathrm{I}\oplus\bar{1}\bar{0}\oplus\bar{5}\bar{4}\bar{1}0\oplus\bar{5}\bar{4}1\bar{0}$	0 1 2 3 4 5	$\mathrm{I}\oplus\bar{1}\bar{0}\oplus\bar{5}\bar{4}\bar{1}0\oplus\bar{5}\bar{4}1\bar{0}$

Table 4.V. (Continuation)

$\underline{x}_0$	$pf/p\underline{x}_0$	$qf/q\underline{x}_0$	$\underline{x}_0$	$pf/p\underline{x}_0$	$qf/q\underline{x}_0$
∅	$I\oplus 10\oplus\bar{5}4\oplus 320$	$I\oplus 10\oplus\bar{5}4\oplus 320$	2 3	$\bar{0}\oplus\bar{5}4\bar{0}$	$I\oplus\bar{5}4\bar{0}$
0	$\bar{1}\oplus\bar{5}4\bar{1}\oplus 32\oplus\bar{5}432$	$I\oplus\bar{5}4\bar{1}\oplus 5432$	2 4	$5(\bar{3}10\oplus 30\oplus I)$	$I\oplus 5\bar{3}10$
1	$I\oplus 0\oplus\bar{5}4\bar{0}$	$I\oplus\bar{5}4\bar{0}$	2 5	$\bar{4}(I\oplus 30\oplus\bar{3}10)$	$I\oplus\bar{4}\bar{3}10$
2	$I\oplus\bar{3}10\oplus\bar{5}4\oplus\bar{5}430\oplus 30$	$I\oplus\bar{3}10\oplus\bar{5}4\oplus\bar{5}430$	3 4	$5(I\oplus 20\oplus\bar{2}10)$	$I\oplus 5\bar{2}10$
3	$I\oplus\bar{2}10\oplus\bar{5}4\oplus\bar{5}420\oplus 20$	$I\oplus\bar{2}10\oplus\bar{5}4\oplus\bar{5}420$	3 5	$\bar{4}(I\oplus 20\oplus\bar{2}10)$	$I\oplus\bar{4}\bar{2}10$
4	$5\oplus 510\oplus 5320$	$I\oplus 510\oplus 5320$	4 5	∅	I
5	$I\oplus\bar{4}10\oplus 4\oplus\bar{4}320$	$I\oplus\bar{4}10\oplus\bar{4}320$	0 1 2	∅	I
0 1	∅	I	0 1 3	∅	I
0 2	$\bar{3}\bar{1}\oplus\bar{5}4\bar{3}\bar{1}$	$I\oplus\bar{5}4\bar{3}\bar{1}$	0 1 4	∅	I
0 3	$I\oplus\bar{5}4\oplus\bar{2}1\oplus\bar{5}4\bar{2}1\oplus\bar{5}42\oplus 2$	$I\oplus\bar{5}4\oplus\bar{5}421\oplus\bar{5}42$	0 1 5	∅	I
0 4	$5\bar{1}\oplus 532$	I	0 2 3	∅	I
0 5	$\bar{4}\bar{1}\oplus\bar{4}32$	I	0 2 4	$5\bar{3}\bar{1}$	I
1 2	$\bar{0}\oplus\bar{5}4\bar{0}$	$I\oplus\bar{5}4\bar{0}$	0 2 5	$\bar{4}\bar{3}\bar{1}$	I
1 3	$\bar{0}\oplus\bar{5}4\bar{0}$	$I\oplus\bar{5}4\bar{0}$	0 3 4	$5\bar{2}\bar{1}$	I
1 4	$\bar{0}5$	I	0 3 5	$\bar{4}\bar{2}\bar{1}$	I
1 5	$\bar{0}\bar{4}$	I	0 4 5	∅	I

Table 4. VI.

$\underline{x}_0$	$pf/p\underline{x}_0$	$qf/q\underline{x}_0$	$\underline{x}_0$	$pf/p\underline{x}_0$	$qf/q\underline{x}_0$
1 2 3	$\bar{0}\oplus\bar{5}4\bar{0}$	$I\oplus\bar{5}4\bar{0}$	0 2 3 4	Ø	I
1 2 4	$5\bar{0}$	I	0 2 3 5	Ø	I
1 2 5	$\bar{4}\bar{0}$	I	0 2 4 5	Ø	I
1 3 4	$5\bar{0}$	I	0 3 4 5	Ø	I
1 3 5	$\bar{4}\bar{0}$	I	1 2 3 4	$5\bar{0}$	I
1 4 5	Ø	I	1 2 3 5	$\bar{4}\bar{0}$	I
2 3 4	$5\bar{0}$	I	1 2 4 5	Ø	I
2 3 5	$\bar{4}\bar{0}$	I	1 3 4 5	Ø	I
2 4 5	Ø	I	2 3 4 5	Ø	I
3 4 5	Ø	I	0 1 2 3 4	Ø	I
0 1 2 3	Ø	I	0 1 2 3 5	Ø	I
0 1 2 4	Ø	I	0 1 2 4 5	Ø	I
0 1 2 5	Ø	I	0 1 3 4 5	Ø	I
0 1 3 4	Ø	I	0 2 3 4 5	Ø	I
0 1 3 5	Ø	I	1 2 3 4 5	Ø	I
0 1 4 5	Ø	I	0 1 2 3 4 5	Ø	I

Table 4.VI. (Continuation)

$\underline{x}_0$	$\delta f/\delta \underline{x}_0$	$p_{\underline{x}_0} f,\ q_{\underline{x}_0} f$	$\underline{x}_0$	$\delta f/\delta \underline{x}_0$	$p_{\underline{x}_0} f,\ q_{\underline{x}_0} f$
∅	∅	$0\bar{1}234\bar{5}$	2 3	0	∅
0	$1\oplus 32$	$\underline{1234\bar{5}}$	2 4	$\bar{5}\vee 30$	$\underline{01\bar{3}5}$
1	0	∅	2 5	$4\vee 30$	$\underline{01\bar{3}\bar{4}}$
2	30	$01\bar{3}4\bar{5}$	3 4	$\bar{5}\vee 20$	$\underline{01\bar{2}5}$
3	20	$01\bar{2}4\bar{5}$	3 5	$4\vee 20$	$\underline{01\bar{2}\bar{4}}$
4	$\bar{5}$	$\underline{0\bar{1}235}$	4 5	I	∅
5	4	$\underline{0\bar{1}23\bar{4}}$	0 1 2	I	∅
0 1	I	∅	0 1 3	I	∅
0 2	$3\vee 1$	$\underline{\bar{1}\bar{3}4\bar{5}}$	0 1 4	I	∅
0 3	$2\vee 1$	$\underline{\bar{1}\bar{2}4\bar{5}}$	0 1 5	I	∅
0 4	$\bar{5}\vee(1\oplus 32)$	1235	0 2 3	I	∅
0 5	$4\vee(1\oplus 32)$	$123\bar{4}$	0 2 4	$\bar{5}\vee 3\vee 1$	$5\bar{3}\bar{1}$
1 2	0	∅	0 2 5	$4\vee 3\vee 1$	$\bar{4}\bar{3}\bar{1}$
1 3	0	∅	0 3 4	$\bar{5}\vee 2\vee 1$	$5\bar{2}\bar{1}$
1 4	$0\vee\bar{5}$	∅	0 3 5	$4\vee 2\vee 1$	$\bar{4}\bar{2}\bar{1}$
1 5	$0\vee 4$	∅	0 4 5	I	∅

Table 4.VII.

$\underline{x}_0$	$\delta f/\delta \underline{x}_0$	$p_{\underline{x}_0} f,\ q_{\underline{x}_0} f$	$\underline{x}_0$	$\delta f/\delta \underline{x}_0$	$p_{\underline{x}_0} f,\ q_{\underline{x}_0} f$
1 2 3	0	$\underline{\bar{5}4\bar{0}}$	0 2 3 4	I	∅
1 2 4	$\bar{5}$v0	∅	0 2 3 5	I	∅
1 2 5	4v0	∅	0 2 4 5	I	∅
1 3 4	$\bar{5}$v0	∅	0 3 4 5	I	∅
1 3 5	4v0	∅	1 2 3 4	$\bar{5}$v0	$5\bar{0}$
1 4 5	I	∅	1 2 3 5	4v0	$\bar{4}\bar{0}$
2 3 4	$\bar{5}$v0	∅	1 2 4 5	I	∅
2 3 5	4v0	∅	1 3 4 5	I	∅
2 4 5	I	∅	2 3 4 5	I	∅
3 4 5	I	∅	0 1 2 3 4	I	∅
0 1 2 3	I	∅	0 1 2 3 5	I	∅
0 1 2 4	I	∅	0 1 2 4 5	I	∅
0 1 2 5	I	∅	0 1 3 4 5	I	∅
0 1 3 4	I	∅	0 2 3 4 5	I	∅
0 1 3 5	I	∅	1 2 3 4 5	I	∅
0 1 4 5	I	∅	0 1 2 3 4 5	I	∅

Table 4.VII. (Continuation)

Finally the functions $p_{\underline{x}_0} f$ and $q_{\underline{x}_0} f$ (see theorem 4.3.1.3.) obtained from $pf/p\underline{x}_0$ and from $qf/q\underline{x}_0$ are gathered in Table 4.VII. The computation of the functions $q_{\underline{x}_0} f$ (these functions have been underlined in the entries of Table 4.VII in order to distinguish them from the functions $p_{\underline{x}_0} f$) allows us to obtain the expression of f as a conjunction of all its prime implicates, i.e. :

$$\begin{aligned} f=&(\bar{x}_1 \vee \bar{x}_2 \vee \bar{x}_3 \vee \bar{x}_4 \vee x_5)(\bar{x}_0 \vee x_1 \vee \bar{x}_2 \vee \bar{x}_3 \vee \bar{x}_5)(\bar{x}_0 \vee x_1 \vee \bar{x}_2 \vee \bar{x}_3 \vee x_4)(x_1 \vee x_3 \vee \bar{x}_4 \vee x_5)(x_1 \vee x_2 \vee \bar{x}_4 \vee x_5) \\ &(\bar{x}_0 \vee \bar{x}_1 \vee x_3 \vee \bar{x}_5)(\bar{x}_0 \vee \bar{x}_1 \vee x_3 \vee x_4)(\bar{x}_0 \vee \bar{x}_1 \vee x_2 \vee \bar{x}_5)(\bar{x}_0 \vee \bar{x}_1 \vee x_2 \vee x_4)(x_0 \vee \bar{x}_4 \vee x_5) \ . \end{aligned} \tag{4.29}$$

It is indeed clear that all the algorithmspresented in the above sections have a dual statement allowing us to detect all prime implicates of the function f.

4.3.1.8. Algorithm grounded on the use of the extended vector.

The algorithm based on the use of the extended vector and developed in section 4.2.1. can be significantly simplified by use of the following remarks. Let us first introduce some new notations.

We restrict ourselves to the extended vector $\underline{\phi}^{(\wedge)}(f)$ which leads to the obtention of the prime implicants of f ; similar kinds of simplifications for the other types of extended vectors could also be developed.

In the extended vector algorithm, the computation of figure 4.1. leads first to the obtention of the extended vector $\underline{\phi}^{(\wedge)}(f)$ which is a vector of Boolean constants. If the computation is stopped before the obtention of the elements of the extended vector, the result is a vector of Boolean functions. Let us introduce a labelling for the tree-like computation scheme of figure 4.1 ; the three arrows starting from f are labelled with the symbols $(\overline{n-1})$ $(n-1)$ and no symbol respectively. These arrows correspond to the functions $f(x_{n-1}=0)$, $f(x_{n-1}=1)$ and $f(x_{n-1}=0)\ f(x_{n-1}=1)$ respectively. The same scheme is then adopted for each of the above functions but with respect to the variable x_{n-2} ; an iterative labelling process is thus built with respect to the successive variables of $\underline{x}=(x_{n-1},x_{n-2},\ldots,x_1,x_0)$.

<u>R1.</u> During the research of the prime implicants, each branch of the algorithm (as depicted by figure 4.1) may be ended once a cube of the form :

$$\dot{x}_i \wedge \dot{x}_j \wedge \ldots \wedge \dot{x}_k \qquad (\dot{x}=x \text{ or } \bar{x})$$

has been obtained.

If the set of consecutive arrows from f until that implicant are labelled with symbols

$$\dot{p},\dot{q},\ldots,\dot{r} \qquad (\dot{p}=p \text{ or } \bar{p})$$

then the cube

$$\dot{x}_i \wedge \dot{x}_j \wedge \ldots \wedge \dot{x}_k \wedge \dot{x}_p \wedge \dot{x}_q \wedge \ldots \wedge \dot{x}_r$$

is an implicant of f. All implicants which could be obtained by continuing the initial proposed algorithm once a cube has been reached are smaller than or equal to the implicant of f noted above.

The first rule proposed for simplifying the algorithm is to recognize as soon as possible the cubes during the obtention of the treelike scheme of figure 4.1. This allows us to eliminate at this point all the subsequent computations.

R2. If, at each step of the algorithm, the arrows with the fewest number of labels are first explored, the prime implicants are detected in decreasing order of their dimension, so that those parts of the algorithm which would lead to smaller implicants could easily be detected and would thus be ended at once.

R3. Assume that after a set of consecutive arrows labelled

$$\dot{p}, \dot{q}, \ldots, \dot{r}$$

a function g is obtained and assume moreover that this part of the algorithm issued from the continuation of the computation after the obtention of g gives the m prime implicants of f:

$$\{(\dot{x}_p \wedge \dot{x}_q \wedge \ldots \wedge \dot{x}_r \wedge \alpha_0), \ldots, (\dot{x}_p \wedge \dot{x}_q \wedge \ldots \wedge \dot{x}_r \wedge \alpha_{n-1})\}$$

with $\alpha_0, \alpha_1, \ldots, \alpha_{m-1}$, arbitrary cubes. Then, if the same function g is obtained in any other part of the algorithm and after a set of consecutive arrows labelled

$$\dot{a}, \dot{b}, \ldots, \dot{c} \ ,$$

the following cubes are the largest implicants of f which could be obtained by developing further on that part of the algorithm which follows the obtention of g :

$$\{(\dot{x}_a \wedge \dot{x}_b \wedge \ldots \wedge \dot{x}_c \wedge \alpha_0), \ldots, (\dot{x}_a \wedge \dot{x}_b \wedge \ldots \wedge \dot{x}_c \wedge \alpha_{m-1})\}.$$

The three above remarks allow us to achieve the initial proposed extended-vector algorithm.

This modified extended vector algorithm applied to the research of the prime implicants of the switching function 4.3.1.7. is illustrated by means of figure 4.3. The prime implicants have been underlined in that figure. Moreover, identical functions appearing at different points of the algorithm (illustration of the simplification rule R3) are designated by asterisks, i.e. $(\ )^{*}$ and $(\ )^{**}$

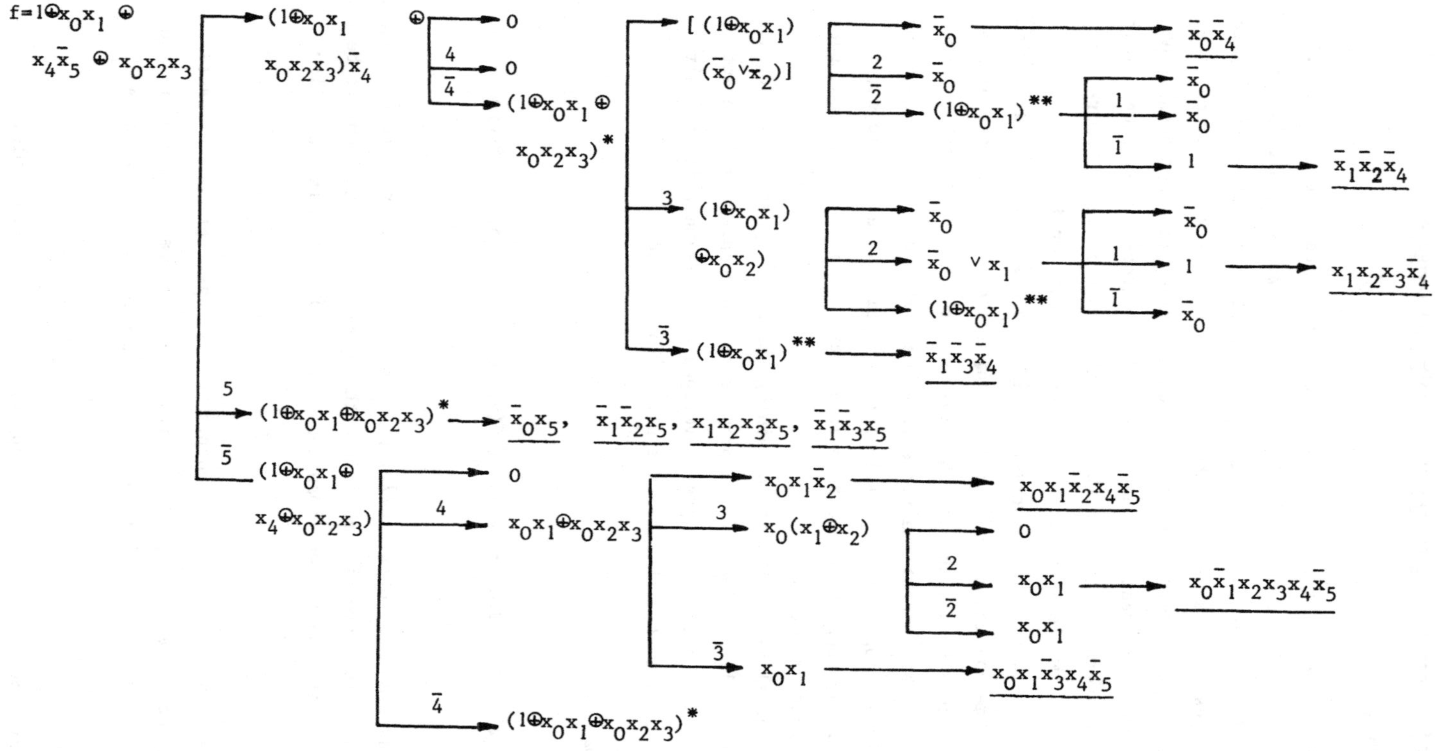

$f=1\oplus x_0x_1 \oplus x_4\bar{x}_5 \oplus x_0x_2x_3$
$(1\oplus x_0x_1 \oplus x_0x_2x_3)\bar{x}_4$
$[(1\oplus x_0x_1)(\bar{x}_0 \vee \bar{x}_2)]$
$(1\oplus x_0x_1 \oplus x_0x_2x_3)^*$
$(1\oplus x_0x_1)^{**}$
$\bar{x}_0 \vee x_1$
$\bar{x}_0\bar{x}_4$
$\bar{x}_1\bar{x}_2\bar{x}_4$
$x_1x_2x_3\bar{x}_4$
$\bar{x}_1\bar{x}_3\bar{x}_4$
$\bar{x}_0x_5$, $\bar{x}_1\bar{x}_2x_5$, $x_1x_2x_3x_5$, $\bar{x}_1\bar{x}_3x_5$
$(1\oplus x_0x_1 \oplus x_4 \oplus x_0x_2x_3)$
$x_0x_1 \oplus x_0x_2x_3$
$x_0x_1\bar{x}_2$
$x_0(x_1\oplus x_2)$
$x_0x_1\bar{x}_2x_4\bar{x}_5$
$x_0\bar{x}_1x_2x_3x_4\bar{x}_5$
$x_0x_1\bar{x}_3x_4\bar{x}_5$

Figure 4.3.

4.3.2. Synthesis of three-level networks using AND- and OR-gates

4.3.2.1. Problem statement

The synthesis of three-level networks is based on theorem 3.2.18. Remember that the synthesis of three-level AND-OR-AND networks is obtained in the following way. The AND-gates of the first level realize the prime implicants of the upper envelopes. The OR-gates of the second level realize the prime upper envelopes (i.e. the upper envelopes which are not contained in other ones). The AND-gate of the third level realizes the function f. As quoted in section 3.2.18, the interest of these networks lies in the fact that they contain no more logic hazards than the two level networks. Let us only mention that Eichelberger [1965] stated that the two-level networks realizing the disjunction of all prime implicants of f or the conjunction of all prime implicates of f are free of static logic hazards (see also section 3.2.18). The author (Thayse [1975b], Thayse and Deschamps [1977]) showed that the three-level networks, built-up as a conjunction of prime upper envelopes or as a disjunction of prime lower envelopes, are free of logic static hazards and contain no more logic dynamic hazards than the two-level Eichelberger networks.

Universal algorithms for obtaining the prime upper and lower envelopes have been stated in section 4.2. We propose here a completely different algorithm allowing us to obtain the upper envelopes as a disjunction of their prime implicants; it is a straightforward application of theorem 3.2.17.

4.3.2.2. Algorithm

Step 1

Starting from any well-formed expression of $f(\underline{x})$, obtain one of its Newton expansions.

Step 2

From a Newton expansion of f, obtain *all* its Newton expansions by proceeding along a Gray-code ordering. This ordering minimizes the number of computations to be performed for obtaining all the Newton expansions (see Davio, Deschamps and Bioul [1974]).

Step 3

The upper envelopes (expressed as a disjunction of all their prime implicants) are obtained from the corresponding Newton expansions by replacing the operation "$\oplus$" by the operation "v" and by deleting the cubes contained in other cubes (absorption rule).

Step 4

The prime upper envelopes are obtained by applying the absorption rule.

4.3.2.3. Example

We continue example 4.3.1.7.
From the Newton expansion of f :

$$1 \oplus x_1x_0 \oplus x_4 \oplus x_5x_4 \oplus x_3x_2x_0$$

We deduce all Newton expansions (see Table 4.VIII). From these Newton expansions we obtain the upper envelopes (see Table 4.IX).
We deduce finally the expression of f as a conjunction of its prime upper envelopes, each of these envelopes being expressed as a disjunction of its prime implicants, i.e. :

$$\begin{aligned} f = & (\bar{x}_4 \vee x_5 \vee \bar{x}_1x_0 \vee \bar{x}_3x_0 \vee \bar{x}_2x_0)(\bar{x}_4 \vee x_5 \vee x_1x_0 \vee x_3x_2x_0) \\ & (x_1 \vee \bar{x}_0 \vee \bar{x}_2 \vee \bar{x}_3 \vee \bar{x}_5x_4)(\bar{x}_1 \vee \bar{x}_0 \vee \bar{x}_5x_4 \vee x_3x_2) \end{aligned} \quad (4.30)$$

4.3.2.3. Some further considerations related to logical design

Until now we have taken into account two important parameters in logical design, namely the *cost*, i.e. the number of gates present in the network, and the possibility of presence of logic hazards. Let us point out that the minimization of the number of gates and of the number of logic hazards constitute generally two incompatible goals. We know e.g. that the elimination of all the logic static hazards requests the presence of as many AND-gates as prime implicants in a two-level AND-OR network while the minimization of the cost requests as many AND-gates as the number of prime implicants in an irredundant cover of f. In this respect, consider e.g. the two- and three-level realizations of the function (4.28). Since this function has 11 and 10 prime implicants and prime implicates respectively the corresponding two level AND-OR and OR-AND networks (free of static-logic hazards) have 12 and 11 gates respectively. The three level realization corresponding to the formula (4.30) has 13 gates. We know that three-level realizations eliminate some dynamic logic hazards occuring in two-level realizations.

A new practical limitation to be introduced in the design of switching networks is the FAN-IN. Diode gates can be extended to generate AND or OR gates of more than two variables, but such an extension cannot be continued without limit. Diodes are not ideal : they have nonzero forward resistance and finite reverse resistance for example.
If too many diodes are included in a gate that gate will not operate reliably. The maximum fan-in of such gates is limited. The diode characteristics determine the exact limit but 10 is a typical limit. Thus the conjunction of 11 literals may

$x_5x_4x_3x_2x_1x_0$	W(f)		$x_5x_4x_3x_2x_1x_0$	W(f)	
0 0 0 0 0 0	5	$I\oplus 10\oplus 4\oplus 54\oplus 320$	0 1 1 0 0 0	6	$10\oplus\bar{4}\oplus 5\oplus 5\bar{4}\oplus\bar{3}20\oplus 20$
0 0 0 0 0 1	7	$I\oplus 1\oplus 1\bar{0}\oplus 4\oplus 54\oplus 32\oplus 32\bar{0}$	0 1 1 0 0 1	9	$1\oplus 1\bar{0}\oplus\bar{4}\oplus 5\oplus 5\bar{4}\oplus\bar{3}2\oplus\bar{3}2\bar{0}\oplus 2\oplus 2\bar{0}$
0 0 0 0 1 1	7	$\bar{1}\oplus\bar{0}\oplus\bar{1}\bar{0}\oplus 4\oplus 54\oplus 32\oplus 32\bar{0}$	0 1 1 0 1 1	11	$I\oplus\bar{1}\oplus\bar{0}\oplus\bar{1}\bar{0}\oplus\bar{4}\oplus 5\oplus 5\bar{4}\oplus\bar{3}2\oplus\bar{3}2\bar{0}\oplus 2\oplus 2\bar{0}$
0 0 0 0 1 0	6	$I\oplus 0\oplus\bar{1}0\oplus 4\oplus 54\oplus 320$	0 1 1 0 1 0	7	$0\oplus\bar{1}0\oplus\bar{4}\oplus 5\oplus 5\bar{4}\oplus\bar{3}20\oplus 20$
0 0 0 1 1 0	7	$I\oplus 0\oplus\bar{1}0\oplus 4\oplus 54\oplus 30\oplus 3\bar{2}0$	0 1 1 1 1 0	7	$\bar{1}0\oplus\bar{4}\oplus 5\oplus 5\bar{4}\oplus\bar{3}\bar{2}0\oplus\bar{3}0\oplus\bar{2}0$
0 0 0 1 1 1	9	$\bar{0}\oplus\bar{1}\oplus\bar{1}\bar{0}\oplus 4\oplus 54\oplus 3\oplus 3\bar{0}\oplus 3\bar{2}\oplus 3\bar{2}\bar{0}$	0 1 1 1 1 1	11	$\bar{1}\bar{0}\oplus\bar{1}\oplus\bar{4}\oplus 5\oplus 5\bar{4}\oplus\bar{3}\bar{2}\bar{0}\oplus\bar{3}\bar{2}\oplus\bar{3}\bar{0}\oplus\bar{3}\oplus\bar{2}\bar{0}\oplus\bar{2}$
0 0 0 1 0 1	9	$I\oplus 1\oplus 1\bar{0}\oplus 4\oplus 54\oplus 3\oplus 3\bar{0}\oplus 3\bar{2}\oplus 3\bar{2}\bar{0}$	0 1 1 1 0 1	13	$I\oplus 1\oplus 1\bar{0}\oplus\bar{0}\oplus\bar{4}\oplus 5\oplus 5\bar{4}\oplus\bar{3}\bar{2}\bar{0}\oplus\bar{3}\bar{2}\oplus\bar{3}\bar{0}\oplus\bar{3}\oplus\bar{2}\bar{0}\oplus\bar{2}$
0 0 0 1 0 0	6	$I\oplus 10\oplus 4\oplus 54\oplus 30\oplus 3\bar{2}0$	0 1 1 1 0 0	8	$0\oplus 10\oplus\bar{4}\oplus 5\oplus 5\bar{4}\oplus\bar{3}\bar{2}0\oplus\bar{3}0\oplus\bar{2}0$
0 0 1 1 0 0	8	$I\oplus 10\oplus 4\oplus 54\oplus\bar{3}0\oplus 0\oplus\bar{3}\bar{2}0\oplus\bar{2}0$	0 1 0 1 0 0	6	$10\oplus\bar{4}\oplus 5\oplus 5\bar{4}\oplus 3\bar{2}0\oplus 30$
0 0 1 1 0 1	11	$\bar{0}\oplus 1\bar{0}\oplus 1\oplus 4\oplus 54\oplus\bar{3}\bar{0}\oplus\bar{3}\oplus\bar{3}\bar{2}\bar{0}\oplus\bar{3}\bar{2}\ \oplus\bar{2}\bar{0}\oplus\bar{2}$	0 1 0 1 0 1	9	$1\oplus 1\bar{0}\oplus\bar{4}\oplus 5\oplus 5\bar{4}\oplus 3\bar{2}\oplus 3\bar{2}\bar{0}\oplus 3\oplus 3\bar{0}$
0 0 1 1 1 1	11	$I\oplus\bar{1}\oplus\bar{1}\bar{0}\oplus 4\oplus 54\oplus\bar{3}\bar{0}\oplus\bar{3}\oplus\bar{3}\bar{2}\bar{0}\oplus\bar{3}\bar{2}\oplus\bar{2}\bar{0}\oplus\bar{2}$	0 1 0 1 1 1	11	$I\oplus\bar{1}\oplus\bar{0}\oplus\bar{1}\bar{0}\oplus\bar{4}\oplus 5\oplus 5\bar{4}\oplus 3\bar{2}\oplus 3\bar{2}\bar{0}\oplus 3\oplus 3\bar{0}$
0 0 1 1 1 0	7	$I\oplus\bar{1}0\oplus 4\oplus 54\oplus\bar{3}0\oplus\bar{3}\bar{2}0\oplus\bar{2}0$	0 1 0 1 1 0	7	$0\oplus\bar{1}0\oplus\bar{4}\oplus 5\oplus 5\bar{4}\oplus 3\bar{2}0\oplus 30$
0 0 1 0 1 0	7	$I\oplus\bar{1}0\oplus 4\oplus 54\oplus\bar{3}20\oplus 20\oplus 0$	0 1 0 0 1 0	6	$0\oplus\bar{1}0\oplus\bar{4}\oplus 5\oplus 5\bar{4}\oplus 320$
0 0 1 0 1 1	9	$\bar{1}\oplus\bar{1}\bar{0}\oplus 4\oplus 54\oplus\bar{3}2\oplus\bar{3}2\bar{0}\oplus 2\oplus 2\bar{0}\oplus\bar{0}$	0 1 0 0 1 1	9	$I\oplus\bar{0}\oplus\bar{1}\oplus\bar{1}\bar{0}\oplus\bar{4}\oplus 5\oplus 5\bar{4}\oplus 32\oplus 32\bar{0}$
0 0 1 0 0 1	9	$I\oplus 1\oplus 1\bar{0}\oplus 4\oplus 54\oplus\bar{3}2\oplus\bar{3}2\bar{0}\oplus 2\oplus 2\bar{0}$	0 1 0 0 0 1	7	$1\bar{0}\oplus 1\oplus\bar{4}\oplus 5\oplus 5\bar{4}\oplus 32\oplus 32\bar{0}$
0 0 1 0 0 0	6	$I\oplus 10\oplus 4\oplus 54\oplus\bar{3}20\oplus 20$	0 1 0 0 0 0	5	$10\oplus\bar{4}\oplus 5\oplus 5\bar{4}\oplus 320$

Table 4. VIII.

$x_5x_4x_3x_2x_1x_0$	W(f)		$x_5x_4x_3x_2x_1x_0$	W(f)	
1 1 0 0 0 0	5	$I\oplus\bar{5}\oplus\bar{5}\bar{4}\oplus 10\oplus 320$	1 0 1 0 0 0	5	$I\oplus 10\oplus\bar{5}4\oplus\bar{3}20\oplus 20$
1 1 0 0 0 1	7	$I\oplus 1\oplus 1\bar{0}\oplus\bar{5}\oplus\bar{5}\bar{4}\oplus 32\oplus 32\bar{0}$	1 0 1 0 0 1	8	$I\oplus 1\oplus 1\bar{0}\oplus\bar{5}4\oplus\bar{3}2\oplus\bar{3}2\bar{0}\oplus 2\oplus 2\bar{0}$
1 1 0 0 1 1	7	$\bar{1}\oplus\bar{0}\oplus\bar{1}\bar{0}\oplus\bar{5}\oplus\bar{5}\bar{4}\oplus 32\oplus 32\bar{0}$	1 0 1 0 1 1	8	$\bar{1}\oplus\bar{0}\oplus\bar{1}\bar{0}\oplus\bar{5}4\oplus\bar{3}2\oplus\bar{3}2\bar{0}\oplus 2\oplus 2\bar{0}$
1 1 0 0 1 0	6	$I\oplus 0\oplus\bar{1}0\oplus\bar{5}\oplus\bar{5}\bar{4}\oplus 320$	1 0 1 0 1 0	6	$I\oplus 0\oplus\bar{1}0\oplus\bar{5}4\oplus\bar{3}20\oplus 20$
1 1 0 1 1 0	7	$I\oplus 0\oplus\bar{1}0\oplus\bar{5}\oplus\bar{5}\bar{4}\oplus 30\oplus 3\bar{2}0$	1 0 1 1 1 0	6	$I\oplus\bar{1}0\oplus\bar{5}4\oplus\bar{3}\bar{2}0\oplus\bar{3}0\oplus\bar{2}0$
1 1 0 1 1 1	9	$\bar{0}\oplus\bar{1}\oplus\bar{1}\bar{0}\oplus\bar{5}\oplus\bar{5}\bar{4}\oplus 3\oplus 3\bar{0}\oplus 3\bar{2}\oplus 3\bar{2}\bar{0}$	1 0 1 1 1 1	10	$I\oplus\bar{1}\bar{0}\oplus\bar{1}\oplus\bar{5}4\oplus\bar{3}\bar{2}\bar{0}\oplus\bar{3}\bar{2}\oplus\bar{3}\bar{0}\oplus\bar{3}\oplus\bar{2}\bar{0}\oplus\bar{2}$
1 1 0 1 0 1	9	$I\oplus 1\oplus 1\bar{0}\oplus\bar{5}\oplus\bar{5}\bar{4}\oplus 3\oplus 3\bar{0}\oplus 3\bar{2}\oplus 3\bar{2}\bar{0}$	1 0 1 1 0 1	10	$1\oplus 1\bar{0}\oplus\bar{0}\oplus\bar{5}4\oplus\bar{3}\bar{2}\bar{0}\oplus\bar{3}\bar{2}\oplus\bar{3}\bar{0}\oplus\bar{3}\oplus\bar{2}\bar{0}\oplus\bar{2}$
1 1 0 1 0 0	6	$I\oplus 10\oplus\bar{5}\oplus\bar{5}\bar{4}\oplus 30\oplus 3\bar{2}0$	1 0 1 1 0 0	7	$I\oplus 0\oplus 10\oplus\bar{5}4\oplus\bar{3}\bar{2}0\oplus\bar{3}0\oplus\bar{2}0$
1 1 1 1 0 0	8	$I\oplus 10\oplus\bar{5}\oplus\bar{5}\bar{4}\oplus\bar{3}0\oplus 0\oplus\bar{3}\bar{2}0\oplus\bar{2}0$	1 0 0 1 0 0	5	$I\oplus 10\oplus\bar{5}4\oplus 3\bar{2}0\oplus 30$
1 1 1 1 0 1	11	$\bar{0}\oplus 1\bar{0}\oplus 1\oplus\bar{5}\oplus\bar{5}\bar{4}\oplus\bar{3}\bar{0}\oplus\bar{3}\oplus\bar{3}\bar{2}\bar{0}\oplus\bar{3}\bar{2}\oplus\bar{2}\bar{0}\oplus\bar{2}$	1 0 0 1 0 1	8	$I\oplus 1\oplus 1\bar{0}\oplus\bar{5}4\oplus 3\bar{2}\oplus 3\bar{2}\bar{0}\oplus 3\oplus 3\bar{0}$
1 1 1 1 1 1	11	$I\oplus\bar{1}\oplus\bar{1}\bar{0}\oplus\bar{5}\oplus\bar{5}\bar{4}\oplus\bar{3}\bar{0}\oplus\bar{3}\oplus\bar{3}\bar{2}\bar{0}\oplus\bar{3}\bar{2}\oplus\bar{2}\bar{0}\oplus\bar{2}$	1 0 0 1 1 1	8	$\bar{1}\oplus\bar{0}\oplus\bar{1}\bar{0}\oplus\bar{5}4\oplus 3\bar{2}\oplus 3\bar{2}\bar{0}\oplus 3\oplus 3\bar{0}$
1 1 1 1 1 0	7	$I\oplus\bar{1}0\oplus\bar{5}\oplus\bar{5}\bar{4}\oplus\bar{3}0\oplus\bar{3}\bar{2}0\oplus\bar{2}0$	1 0 0 1 1 0	6	$I\oplus 0\oplus\bar{1}0\oplus\bar{5}4\oplus 3\bar{2}0\oplus 30$
1 1 1 0 1 0	7	$I\oplus\bar{1}0\oplus\bar{5}\oplus\bar{5}\bar{4}\oplus\bar{3}20\oplus 20\oplus 0$	1 0 0 0 1 0	5	$I\oplus 0\oplus\bar{1}0\oplus\bar{5}4\oplus 320$
1 1 1 0 1 1	9	$\bar{1}\oplus\bar{1}\bar{0}\oplus\bar{5}\oplus\bar{5}\bar{4}\oplus\bar{3}2\oplus\bar{3}2\bar{0}\oplus 2\bar{0}\oplus\bar{0}\oplus 2$	1 0 0 0 1 1	6	$\bar{0}\oplus\bar{1}\oplus\bar{1}\bar{0}\oplus\bar{5}4\oplus 32\oplus 32\bar{0}$
1 1 1 0 0 1	9	$I\oplus 1\oplus 1\bar{0}\oplus\bar{5}\oplus\bar{5}\bar{4}\oplus\bar{3}2\oplus\bar{3}2\bar{0}\oplus 2\oplus 2\bar{0}$	1 0 0 0 0 1	6	$I\oplus 1\bar{0}\oplus 1\oplus\bar{5}4\oplus 32\oplus 32\bar{0}$
1 1 1 0 0 0	6	$I\oplus 10\oplus\bar{5}\oplus\bar{5}\bar{4}\oplus\bar{3}20\oplus 20$	1 0 0 0 0 0	4	$I\oplus 10\oplus\bar{5}4\oplus 320$

Table 4.VIII. (Continuation)

$A_5A_4A_3A_2A_1A_0$	$q^{A_5A_4A_3A_2A_1A_0}$ $f/qx_5x_4x_3x_2x_1x_0$	$A_5A_4A_3A_2A_1A_0$	$q^{A_5A_4A_3A_2A_1A_0}$ $f/qx_5x_4x_3x_2x_1x_0$
1 1 0 0 0 0	I	1 0 1 0 0 0	I
1 1 0 0 0 1	I	1 0 1 0 0 1	I
1 1 0 0 1 1	$\bar{1}v\bar{0}v\bar{5}v32$	1 0 1 0 1 1	$\bar{1}v\bar{0}v\bar{5}4v2$
1 1 0 0 1 0	I	1 0 1 0 1 0	I
1 1 0 1 1 0	I	1 0 1 1 1 0	I
1 1 0 1 1 1	$\bar{0}v\bar{1}v\bar{5}v3$	1 0 1 1 1 1	I
1 1 0 1 0 1	I	1 0 1 1 0 1	$1v\bar{0}v\bar{2}v\bar{3}v\bar{5}4$ Prime envelope
1 1 0 1 0 0	I	1 0 1 1 0 0	I
1 1 1 1 0 0	I	1 0 0 1 0 0	I
1 1 1 1 0 1	$\bar{0}v1v\bar{5}v\bar{3}v\bar{2}$	1 0 0 1 0 1	I
1 1 1 1 1 1	I	1 0 0 1 1 1	$\bar{1}v\bar{0}v3v\bar{5}4$
1 1 1 1 1 0	I	1 0 0 1 1 0	I
1 1 1 0 1 0	I	1 0 0 0 1 0	I
1 1 1 0 1 1	$\bar{1}v\bar{5}v\bar{0}v2$	1 0 0 0 1 1	$\bar{0}v\bar{1}v\bar{5}4v32$ Prime envelope
1 1 1 0 0 1	I	1 0 0 0 0 1	I
1 1 1 0 0 0	I	1 0 0 0 0 0	I

Table 4. IX. (Continuation)

$A_5A_4A_3A_2A_1A_0$	$q^{A_5A_4A_3A_2A_1A_0}$ $f/qx_5x_4x_3x_2x_1x_0$	$A_5A_4A_3A_2A_1A_0$	$q^{A_5A_4A_3A_2A_1A_0}$ $f/qx_5x_4x_3x_2x_1x_0$
0 0 0 0 0 0	I	0 1 1 0 0 0	$\bar{4}v5v10v20$
0 0 0 0 0 1	I	0 1 1 0 0 1	$1v\bar{4}v5v2$
0 0 0 0 1 1	$\bar{1}v\bar{0}v4v32$	0 1 1 0 1 1	I
0 0 0 0 1 0	I	0 1 1 0 1 0	$0v\bar{4}v\ 5$
0 0 0 1 1 0	I	0 1 1 1 1 0	$\bar{4}v5v\bar{1}0v\bar{3}0v\bar{2}0$ **Prime envelope**
0 0 0 1 1 1	$\bar{0}v\bar{1}v4v3$	0 1 1 1 1 1	$\bar{1}v\bar{4}v5v\bar{3}v\bar{2}$
0 0 0 1 0 1	I	0 1 1 1 0 1	I
0 0 0 1 0 0	I	0 1 1 1 0 0	$0v\bar{4}v5$
0 0 1 1 0 0	I	0 1 0 1 0 0	$\bar{4}v5v10v30$
0 0 1 1 0 1	$\bar{0}v1v4v\bar{3}v\bar{2}$	0 1 0 1 0 1	$1v\bar{4}v5v3$
0 0 1 1 1 1	I	0 1 0 1 1 1	I
0 0 1 1 1 0	I	0 1 0 1 1 0	$0v\bar{4}v5$
0 0 1 0 1 0	I	0 1 0 0 1 0	$0v\bar{4}v5$
0 0 1 0 1 1	$\bar{1}v4v2v\bar{0}$	0 1 0 0 1 1	I
0 0 1 0 0 1	I	0 1 0 0 0 1	$1v\bar{4}v5v32$
0 0 1 0 0 0	I	0 1 0 0 0 0	$\bar{4}v5v10v320$ **Prime envelope**

Table 4.IX. (Continuation)

not be realizable by a single AND-gate with a fan-in of 10.

Integrated circuit gates must be manufactured with a fixed number of input terminals. Fabrication of such gates has established a fan-in limit, which the designer is required to observe. For example it is classical to integrate OR, AND, NAND and NOR gates having 3 input terminals.

Using the 3-input fan-in constraint, the two-level AND-OR, OR-AND and the three-level realization are transformed into networks containing 22, 24 and 18 gates respectively; the optimal realization corresponds thus in this case to the three-level network.

4.3.3. Two-level networks using AND and EX-OR gates

4.3.3.1. Problem statement

We consider the synthesis of two-level networks, the first level being formed by AND-gates and the second level by an EX-OR gate, the inputs of which being connected to the AND-gates outputs. This is a classical design technique, the properties of which were studied by several authors (Mukhopadhyay and Smith [1970], Bioul and Davio [1972], Reddy [1972], Thayse [1974a]). In particular let us point out that Reddy showed that AND-EX-OR networks are *easily testable networks* ; desirable properties of easily testable networks are briefly recalled below.

Assume that only permanent stuck-at-0 or stuck-at-1 occur in a single AND-gate or in the EX-OR gate ; the AND-EX-OR realization has the following test properties :

- The number of tests to be performed to make sure that the network is fault-free is small ; only (n/4) tests, (n being the number of variables), independent of the function being realized, are required if the network inputs are fault-free. Only $2n_e$ additional tests (which depend on the function realized, are required if the inputs can be faulty, where n_e is the number of variables appearing in an even number of product terms in the Newton (or equivalently Reed-Muller) canonical expansion of the function.

- The test set can be found without much extra work either during the design phase or after the network is designed; the structure is such that the test can be easily generated and the results can be easily interpreted.

Further on we are interested in minimal AND-EX-OR networks, i.e. networks having a minimum number of AND-gates. If we observe that there is a one-to-one correspondence between the number of AND-gates and the number of terms in the Reed-Muller expansion, the problem reduces to finding the vertex (vertices) $\underline{h}$ where a maximum number of Boolean differences vanishes (see theorem 3.2.9).

Several algorithms for solving this problem exist in the literature; in section 4.3.3.2 we state algorithms which are particularly suited when the function f is given by means of a well-formed expression. The algorithm of section 4.3.3.3. is more appropriate when f is given by its truth vector.

4.3.3.2. Algorithms grounded on the use of differential operators

The existence of Newton expansions of Boolean functions is known from the work of Akers. Since then, several authors have raised the problem of finding an expansion having a minimum weight (number of summed terms). These expansions indeed provide some easily testable designs for the Boolean functions. The present section provides us with a method for obtaining a pseudo-Boolean function whose value at any vertex is the weight of the Taylor expansion of the corresponding Boolean function at the same vertex. This method is mainly grounded on some concepts of Boolean difference calculus which were developed in chapters 2,3.

Given a function f of n variables $\underline{x}$, the Newton expansion of f evaluated at the vertex $\underline{h}$ is (see theorem 3.2.9)

$$\bigoplus_{\underline{e}} (\Delta f/\Delta \underline{x}^{\underline{e}})_{\underline{x}=\underline{h}} \bigwedge_{i=0,n-1} (x_i \oplus h_i)^{e_i}, \ 0 \leqslant e_i \leqslant 1, \qquad (4.31)$$

Consequently, the weight of the expansion of f at the point $\underline{h}$ is given by (+ and $\sum$ stand for the real sum) :

$$\sum_{\underline{e}} (\Delta f/\Delta \underline{x}^{\underline{e}})_{\underline{x}=\underline{h}}, \ 0 \leqslant e_i \leqslant 1 \qquad (4.32)$$

In (4.31,32) it is conventionally assumed that the $\underline{0}$-difference $\Delta f/\Delta \underline{x}^{\underline{0}}$ is $f(\underline{x})$. The following algorithm may then be used for building a pseudo-Boolean function w(f) which gives us the weights of the Newton expansions at any vertex and, from that on, the minimal Newton expansions.

1. From any formal representation of $f(\underline{x})$ compute the Boolean differences $\Delta f/\Delta \underline{x}^{\underline{e}}$, $0 \leqslant e_i \leqslant 1$
2. Obtain the expressions of $\Delta f/\Delta \underline{x}^{\underline{e}}$ $\forall$ $\underline{e}$ in terms of real sums by using the following well-known identities :

   $a \oplus b = a+b-2ab$, $a \vee b = a+b-ab$, $\bar{a} = 1-a$,
   Let $\Delta F/\Delta \underline{x}^{\underline{e}}$ be the representation of $\Delta f/\Delta \underline{x}^{\underline{e}}$.
3. Compute the pseudo-Boolean function

$$w(f) = \sum_{\underline{e}} \Delta F/\Delta \underline{x}^{\underline{e}}, \quad 0 \leqslant e_i \leqslant 1 \qquad (4.33)$$

The weight of the Newton expansion of f at the point $\underline{h}$ is then $(w(f))_{\underline{x}=\underline{h}}$.

4. The vertices where w(f) takes its minimal value may then be obtained by using any pseudo-Boolean method (Hammer and Rudeanu [1968] , Rosenberg [1974], Thayse [1979]). Let us illustrate this algorithm by means of a short example. The non-zero differences of the example 3.1.14 are :

Boolean expressions	Real expressions
$f=x_0x_2 \vee x_1\bar{x}_2$	$F = x_0x_2 + x_1 - x_1x_2$
$\Delta f/\Delta x_0 = x_2$	$\Delta F/\Delta x_0 = x_2$
$\Delta f/\Delta x_1 = \bar{x}_2$	$\Delta F/\Delta x_1 = 1-x_2$
$\Delta f/\Delta x_2 = x_0 \oplus x_1$	$\Delta F/\Delta x_2 = x_0 + x_1 - 2x_0x_1$
$\Delta f/\Delta x_0x_2 = \Delta f/\Delta x_1x_2 = 1$	$\Delta F/\Delta x_0x_2 = \Delta F/\Delta x_1x_2 = 1$

$$w(f) = 3+x_0+2x_1+x_0x_2-x_1x_2-2x_0x_1. \tag{4.34}$$

The vertices where w(f) takes its minimal value are :

$h_0h_1h_2=0\ 0\ \times$ ($\times$ = indeterminate value 0 or 1).

The minimal value is 3 and the corresponding Newton expansions are :

$x_0x_2 \oplus x_1 \oplus x_1x_2$ for $h_0h_1h_2 = 0\ 0\ 0$

$x_0 \oplus x_0\bar{x}_2 \oplus x_1\bar{x}_2$ for $h_0h_1h_2 = 0\ 0\ 1$.

We continue example 4.2.1.7.

$$\begin{aligned} f &= 1 \oplus x_1x_0 \oplus x_4 \oplus x_5x_4 \oplus x_3x_2x_0 \\ &= 1 - x_1x_0 - x_4 + x_4x_5 + 2x_4x_1x_0 - 2x_5x_4x_1x_0 - x_3x_2x_0 + 2x_3x_2x_1x_0 \\ &\quad + 2x_4x_3x_2x_0 - 2x_5x_4x_3x_2x_0 - 4x_4x_3x_2x_1x_0 + 4x_5x_4x_3x_2x_1x_0 \ . \end{aligned}$$

$$\frac{\Delta f}{\Delta x_0} = x_1 \oplus x_3x_2 = x_1 + x_3x_2 - 2x_3x_2x_1$$

$$\frac{\Delta f}{\Delta x_1} = x_0 \ ; \ \frac{\Delta f}{\Delta x_2} = x_3x_0 \ ; \ \frac{\Delta f}{\Delta x_3} = x_2x_0 \ ; \ \frac{\Delta f}{\Delta x_4} = \bar{x}_5 = 1-x_5 \ ;$$

$$\frac{\Delta f}{\Delta x_5} = x_4 \ ; \ \frac{\Delta f}{\Delta x_0x_2} = x_3 \ ; \ \frac{\Delta f}{\Delta x_0x_3} = x_2 \ ; \ \frac{\Delta f}{\Delta x_2x_3} = x_0 \ ;$$

$$\frac{\Delta f}{\Delta x_0x_1} = \frac{\Delta f}{\Delta x_4x_5} = \frac{\Delta f}{\Delta x_0x_2x_3} = 1 \ .$$

$$w(f) = f + \sum_{\underline{e}} \frac{\Delta f}{\Delta \underline{x}^{\underline{e}}}$$

$$= 5 + 2x_0+x_1+x_2+x_3-x_5-x_1x_0+x_2x_0+x_3x_0+x_3x_2+x_5x_4$$
$$-2x_3x_2x_1-x_3x_2x_0+2x_3x_2x_1x_0+2x_4x_1x_0-2x_5x_4x_1x_0$$
$$+2x_4x_3x_2x_0-2x_5x_4x_3x_2x_0-4x_4x_3x_2x_1x_0+4x_5x_4x_3x_2x_1x_0 \qquad (4.35)$$

We verify that the function w(f) takes its minimum value at the vertex $(x_5x_4x_3x_2x_1x_0)=(1\ 0\ 0\ 0\ 0\ 0)$; this minimum is 4 (the function w(f) has been written down in table 4.VIII). The optimal Newton expansion is :

$$f = 1 \oplus x_1x_0 \oplus \bar{x}_5x_4 \oplus x_3x_2x_0 \qquad (4.36)$$

4.3.3.3. Algorithm grounded on the use of the Taylor expansions

The above algorithm may be improved by making use of several kinds of simplifications. First of all let us observe that the differences $\Delta F/\Delta \underline{x}^{\underline{e}}$, $e > 0$, may be computed by starting from the expression of the 0-difference F without resorting to a first evaluation of the Boolean differences $\Delta f/\Delta \underline{x}^{\underline{e}}$. Indeed, since :

$$\Delta f/\Delta x_0 = f(x_0{=}0) \oplus f(x_0{=}1) = F(x_0{=}0) \oplus F(x_0{=}1) \qquad (4.37)$$

$$\Delta F/\Delta x_0 = F(x_0{=}0) + F(x_0{=}1) -2\, F(x_0{=}0)\, F(x_0{=}1) \qquad (4.38)$$

one has immediately :

$$\Delta F/\Delta x_0 = (F(x_0{=}1)-F(x_0{=}0))^2 = (dF/dx_0)^2 \qquad (4.39)$$

where dF/dx_0 means the derivative (in the sense of the classical differential calculus) of the function F with respect to x_0. The expressions $\Delta F/\Delta \underline{x}^{\underline{e}}$ are then computed iteratively by using relation (4.39). Since the number of Boolean derivatives of f is 2^n the amount of computations increases quickly with the number of variables, except for those functions many multiple differences of which vanish. The evaluation of the differences may however be avoided by making use of some well-known expansions of numerical analysis. First of all, the classical Taylor formula allows us to write :

$$F(x_0+1, x_1+1, \ldots, x_{n-1}+1) = \sum_{\underline{e}} dF/d\underline{x}^{\underline{e}}, \quad 0 \leq e_i \leq 1. \qquad (4.40)$$

It follows that $w(f)=F(x_0+1,x_1+1,\ldots,x_{n-1}+1)$ for the class of functions f such that $dF/d\underline{x}^{\underline{e}}=\Delta F/\Delta \underline{x}^{\underline{e}}\ \forall\ \underline{e}$. In view of (4.37-39) this class of functions reduces to the increasing unate functions whose differences are also all increasing unate functions in the variables domain {0,1}. These functions are the cubes $\wedge_i x_i$, and thus one has trivially :

$$w(\wedge_i x_i) = \prod_i (x_i+1) \tag{4.41}$$

Let us further observe that the cubes $\wedge_i x_i^{(e_i)}$ (e_i=0,1) are increasing unate functions in the variables $x_i^{(e_i)}$ and one has :

$$w(\wedge_i x_i^{(e_i)}) = \prod_i (x_i^{(e_i)}+1) \tag{4.42}$$

Consider now a function f given as a modulo-2 sum of m cubes, that is :

$$f(\underline{x}) = \bigoplus_{j=0,m-1} [\bigwedge_{x_i \in \underline{x}_j} x_i^{(e_{ij})}] , \underline{x}_j \subseteq \underline{x} \; \forall j \tag{4.43}$$

For each subset of q cubes of f ($2 \leqslant q \leqslant m$) let us define the subsets of $\underline{x}$: $\underline{x}_\alpha$, $\underline{x}_\beta$, $\underline{x}_\gamma$ as follows :

$\underline{x}_\gamma$ is a subset of $\underline{x}$ composed of literals appearing with both polarities in the q cubes considered.

If each of the q cubes includes all the literals of $\underline{x}_\gamma$, then $\underline{x}_\alpha$ and $\underline{x}_\beta$ are defined as follows :

$\underline{x}_\alpha$ is a subset of $\underline{x}$ composed of literals appearing with the same polarity in all the q cubes ;

$\underline{x}_\beta$ is a subset of $\underline{x}$ composed of literals appearing with the same polarity in a (proper) subset of the q cubes.

If some of the q cubes include only a subset of the literals of $\underline{x}_\gamma$, then $\underline{x}_\alpha$ and $\underline{x}_\beta$ are not defined.

The following algorithm may be used for obtaining the weights of the Newton expansions of a function given as a modulo-2 sum of cubes :

1. For each cube evaluate $\prod_i (x_i^{(e_{ij})}+1)$
2. For each subset of q cubes ($2 \leqslant q \leqslant m$) evaluate

$$(-2)^{q-1} \prod_{x_k \in \underline{x}_\alpha} (x_k^{(e_{kj})}+1) \prod_{x_\ell \in \underline{x}_\beta} x_\ell^{(e_{\ell j})}$$

(It is recalled that, by definition of $\underline{x}_\alpha$ and $\underline{x}_\beta$, the e_{kj} and $e_{\ell j}$ do not depend on j).

3. The function w(f) is the sum of the functions evaluated under 1 and 2.

Let us briefly indicate a method of proof for this algorithm. For a function $f(\underline{x})$ given as a modulo-2 sum of two cubes, that is :

$$f(\underline{x}) = \bigwedge_{x_i \in \underline{x}_0} x_i^{(e_{i0})} \oplus \bigwedge_{x_i \in \underline{x}_1} x_i^{(e_{i1})}$$

we have $w(f) \leqslant \prod_i (x_i^{(e_{i0})}+1) + \prod_j (x_j^{(e_{i1})}+1)$ since the Newton expansions of the two above cubes contain generally identical terms which disappear by modulo-2 addition. It could then easily be shown that the contribution of these terms is taken into account by the quantity :

$$-2 \prod_{x_k \in \underline{x}_\alpha} (x_k^{(e_{k0})}+1) \prod_{x_\ell \in \underline{x}_\beta} x_\ell^{(e_{\ell 0})}$$

The general formula for m cubes is derived by induction on the number of cubes. Consider first the example 3.1.14 ; we have :

$$f = x_0x_2 \vee x_1\bar{x}_2$$

$$x(f)= (x_0+1)(x_2+1)+(x_1+1)(2-x_2)-2x_0x_1$$

Consider now a somewhat more elaborated example :

$$f = x_0x_3x_4 \oplus \bar{x}_1\bar{x}_2\bar{x}_3x_4 \oplus \bar{x}_0x_1x_2x_3x_4$$

$$w(f) = [(1+x_0)(1+x_3)+(2-x_1)(2-x_2)(2-x_3)+(2-x_0)(1+x_1)(1+x_2)(1+x_3)-2(1-x_0)$$
$$-2x_0(1-x_1)(1-x_2)-2x_1x_2(1+x_3)] \ (1+x_4)$$

Minimizing vertex : $h_0h_1h_2h_3h_4$ = 11100, minimal value : 6

Minimal expansion :

$$x_4x_3 \oplus \bar{x}_1\bar{x}_2x_4 \oplus \bar{x}_1\bar{x}_2x_3x_4 \oplus \bar{x}_0\bar{x}_1x_3x_4 \oplus \bar{x}_0\bar{x}_2x_3x_4 \oplus \bar{x}_0\bar{x}_1\bar{x}_2x_3x_4$$

The above method is particularly attractive for the functions whose expression contains a small number of cubes. The computations are also easy to handle for functions which are given through the intermediate of a Taylor expansion since for this case the subset $\underline{x}_\gamma$ is always empty.

4.3.3.4. Algorithm grounded on the Kronecker matrix product.

From theorem 2.3.5. we deduce that the function w(f) is given by the formula :

$$w(f) = \underline{\phi}^{(\oplus)}(f) \ [+ \times] \ \left(\bigotimes_{i=n-1,0} \begin{bmatrix} 1 & 0 \\ 0 & 1 \\ 1 & 1 \end{bmatrix} \right) .$$

Knowing the vertices where w(f) reaches its minimum value, we may compute the Newton expansions at these vertices.

4.4. Analysis of combinatorial networks.

4.4.1. Problem statement

The purpose of the analysis of switching networks is to verify that a given network computes effectively the Boolean function it is assumed to realize. In switching theory we distinguish between two types of failures that we must be able to detect during any verification process. The first type of failure is the *permanent failure* or *stuck-at-fault*. This kind of failure is essentially produced by some permanent electrical faults, or physical defects of one or more components of the network. Many failures in electrical circuits create a stuck-at-fault in the corresponding logical network model, i.e. produce input or internal variables permanently fixed at the values 0 or 1. A second kind of failure is the *transient failure* or *hazard*. This kind of failure is due to the presence of *delays* in the gates and electrical connections. Failures produced by delays are essentially transient faults since they occur only when the network is working, i.e. when its input variables are changing.

4.4.2. Hazard detection

4.4.2.1. Problem statement

Hazard detection has already been introduced in section 3.2.8. Hazard detection and correction in logical systems is a classical subject in switching theory. The first authors who have stated the problem are Huffman [1957], Unger [1959] and Mc Cluskey [1965].

Binary circuits and the use of the binary number system constitutes the most convenient method of representing information in electronic digital computers and also in many other types of digital systems for control or switching processes. Actual methods for analysis and synthesis of logical networks almost exclusively rely on the use of two-valued Boolean algebra ; this is mainly due to the fact that ideal basic electronic gates realize the operations of two-valued Boolean algebra (such as AND, OR and NOT). However two-valued Boolean algebra is a correct representation of logic networks only for static conditions, i.e. when the signals in the circuit do not change. Whenever the input signals of a combinational or sequential switching network are changed, i.e. for transient conditions, the use of the truth table associated with a given network and deduced from the properties of two-valued Boolean algebra can lead to an incorrect prediction of the real behavior of the network ; the network is then said to contain a *hazard* for that input change. The specific reason for this incorrect prediction is that two-valued Boolean algebra inherently implies the assumption that the propagation time of signals in the logical gates and in the wires interconnecting these logical gates is strictly zero. However no matter what types of real logical gates we use to build a network, there will be an inherent time delay associated with the operation of each of the physical gates. Moreover delays of signal propagation occur for signals to be transmitted along the wires interconnecting the logical gates of a network. The new degrees of freedom brought into the network make that the two-valued Boolean algebra may no longer be used for the analysis of logical networks under transient conditions. Due to the fact that gates and wires exhibit unplanned delays, generally called *stray delays*, transient output values of switching networks may differ from their final values. If it is possible (for at least one combination of the values of the stray delays) for the network output signal to behave in a way different from that predicted by the truth table of the switching function, we say that the network possesses a hazard for the given input transition.

Otherwise stated, transient errors in a network output due to the presence of stray delays are the result of a hazard. Networks in which such transient errors may occur for some distribution of stray delays are said to have one or more hazards. Note that hazards are associated with network configurations, not with physical circuits. A particular physical circuit corresponding to a configuration with a hazard may or may not malfunction depending upon the magnitudes and locations of its stray delays at a particular time. A *hazard-free network* is one that does not display the type of malfunction under discussion, regardless of the distribution of the stray delays.

Hazards are classified in different ways ; according to the kind of the network output signal, there are two types of hazards : the *static hazard* and the *dynamic hazard*.

(a) The characteristic of the static hazard is that it causes a transition in an output which is required to remain constant, during a given input variable change.

(b) The *dynamic hazard*, which can occur when the network output is meant to change, causes the output to change three or more times instead of only once.

Hazards are also classified according to whether they can or cannot be detected by means of tests on the switching function to be realized.

(a) *Function hazards* can be detected by performing tests on the switching function (realized by the switching network) while *logic hazards* cannot.

This last classification has been introduced by Eichelberger[1965] . Function hazards are due to stray delays present at the network inputs ; logic hazards are closely related to internal propagation times on the signal paths, i.e. on the internal stray delays. Logic hazards could thus also be defined as those the effects of which may only be made obvious by inserting stray delays in the network internal wiring.

Only the detection of function hazards will be considered in this section.

4.4.2.2. Algorithms

In view of section 3.2.14 the evaluation of algorithms for detecting either static hazards or dynamic hazards reduces to the tabulation of the function $Sf/S\underline{x}_0 \oplus \delta f/\delta\underline{x}_0$. The functions $Sf/S\underline{x}_0$ and $\delta f/\delta\underline{x}_0$ are both obtained e.g. from the Boolean differences, i.e. :

$$Sf/S\underline{x}_0 = \bigoplus \Delta f/\Delta\underline{x}_0^{\underline{e}_0}, \quad \delta f/\delta\underline{x}_0 = \vee \Delta f/\Delta\underline{x}_0^{\underline{e}_0} \quad .$$

Observe also that the functions $\delta f/\delta\underline{x}^{\underline{e}}$ are easily obtained when all prime implicants and all prime implicates of the function f are known. Besides this, the definition of the function $Sf/S\underline{x}_0$, i.e. :

$$Sf/S\underline{x}_0 = f \oplus f(\bar{\underline{x}}_0)$$

constitutes also a simple computation formula.

In summary, we verify that detecting function hazards reduces to obtain the vertices $\underline{x}=\underline{a}$ satisfying the relations :

$$\left(\frac{Sf}{S\underline{x}_0} \oplus \frac{\delta f}{\delta\underline{x}_0}\right)_{\underline{x}=\underline{a}} = 1 \text{ , (detection of static hazards)} \tag{4.44}$$

$$\frac{Sf}{S\underline{x}_0}\left[\bigvee_{\underline{e}_0}\left(\frac{Sf}{S\underline{x}_0^{\underline{e}_0}} \oplus \frac{\delta f}{\delta\underline{x}_0^{\underline{e}_0}}\right)\right]_{\underline{x}=\underline{a}} = 1 \text{ , (detection of dynamic hazards)} \tag{4.45}$$

$$\left[\bigvee_{\underline{e}_0}\left(\frac{Sf}{S\underline{x}_0^{\underline{e}_0}} \oplus \frac{\delta f}{\delta\underline{x}_0^{\underline{e}_0}}\right)\right]_{\underline{x}=\underline{a}} = 1 \text{ (detection of both static and dynamic hazards)} \tag{4.46}$$

Observe also that the concept of hazard is connected to that of degeneracy and of unateness and thus to that of $\underline{A}$-degeneracy (defined in section 3.2.6). A function which is degenerate in a certain domain is free of hazards for any transition in this domain.If a function is $\{\underline{h}_0\}$-degenerate ($h_i \in \{0,1\}$) in a domain, any transition implying a change of variables of $\underline{x}_0$ from $\underline{x}_\varepsilon = \underline{h}_\varepsilon$ to $\underline{x}_\varepsilon = \bar{\underline{h}}_\varepsilon$ (or conversely) is hazard-free ($\underline{x}_\varepsilon \subseteq \underline{x}_0$, $\underline{h}_\varepsilon \subseteq \underline{h}_0$). It follows that the verification of any one of the three identities of theorem 3.2.16 may be used for building algorithms detecting hazards (see also Beister [1974]). Observe also that, once the functions $Sf/S\underline{x}_0$ and $\delta f/\delta\underline{x}_0$ have been evaluated, their modulo-2 sum provides us with the information necessary to detect any function hazard arising during a transition involving the change of the variables $\underline{x}_0$: one has then to evaluate the same function for different vertices. It follows that the function $(Sf/S\underline{x}_0 \oplus \delta f/\delta\underline{x}_0)$ contains in a relatively condensed form the information necessary to detect hazards for a large number of transitions.

4.4.2.3. Example

We continue the example 4.3.1.7.
The functions $\delta f/\delta \underline{x}_0$ (computed from the functions $pf/p\underline{x}_0$ and $qf/q\underline{x}_0$) have been tabulated in Table 4.VII. We compute in Table 4.X. the function $Sf/S\underline{x}_0 \oplus \delta f/\delta \underline{x}_0$ for $\underline{x}_0$ = any subset of $\underline{x}$ containing 2 or 3 variables. These functions allow us to detect the static function hazards for any transition implying a change of 2 or 3 variables.
Note also that the detection of dynamic hazards by application of formula (4.46) requests much more computation than the detection of static hazards. If we want e.g. to detect the dynamic hazards occurring during a transition implying the variation of x_0, x_1 and x_2, we have to evaluate the function :

$$\frac{Sf}{Sx_0x_1x_2}\left[\left(\frac{Sf}{Sx_0x_1} \oplus \frac{\delta f}{\delta x_0x_1}\right) \vee \left(\frac{Sf}{Sx_0x_2} \oplus \frac{\delta f}{\delta x_0x_2}\right) \vee \left(\frac{Sf}{Sx_1x_2} \oplus \frac{\delta f}{\delta x_1x_2}\right)\right] =$$

$$x_3(\bar{x}_1x_2 \vee x_1\bar{x}_2)$$

These transitions present a dynamic hazard for either $x_1x_2x_3$=0 1 1, or $x_1x_2x_3$=1 0 1.

4.4.3. Fault detection

4.4.3.1. Problem statement

We briefly introduced the problem of fault detection in section 3.2.13. The main reasons for which the fault detection theory may easily be stated in terms of Boolean differences lie in the statement of theorem 3.2.13. Further on, we consider only stuck-at faults. Indeed, many faults in electrical circuits create a stuck-at-fault in the corresponding logical network model : in this kind of fault model it is assumed that any electrical fault (such as e.g. short of open circuited diode, broken wire between gates, etc.) can be modelled by a number or connections in the corresponding logical network permanently fixed at a given logical level (0 or 1). The main purpose of network diagnosis is to detect and/or to locate faults in switching networks.

Consider a logic network realizing a logic function f, assume further that the network undergoing a specific fault realizes the logic function $g(\underline{x})$ instead of $f(\underline{x})$. The function

$$t(\underline{x}) = f(\underline{x}) \oplus g(\underline{x}) .$$

was called test function (see 3.2.13) ; any n-tuple $\underline{a}$ of particular values of the input variables $\underline{x}$ is called test vector for this specific fault iff

$\underline{x}_0$	$Sf/S\underline{x}_0 \oplus \delta f/\delta\underline{x}_0$	$\underline{x}_0$	$Sf/S\underline{x}_0 \oplus \delta f/\delta\underline{x}_0$
0 1	$0 \oplus 1 \oplus 23$	0 1 5	$0 \oplus 1 \oplus 4 \oplus 23$
0 2	$3(0 \oplus 1 \oplus 2)$	0 2 3	$\bar{1} \oplus 023 \oplus \bar{0}\bar{2}\bar{3}$
0 3	$2(0 \oplus 1 \oplus 3)$	0 2 4	$3(0 \oplus 1 \oplus 2) \oplus \bar{5}(1 \vee 3)$
0 4	$\bar{5}(1 \oplus 23)$	0 2 5	$3(0 \oplus 1 \oplus 2) \oplus 4(1 \vee 3)$
0 5	$4(1 \oplus 23)$	0 3 4	$2(0 \oplus 1 \oplus 3) \oplus \bar{5}(1 \vee 2)$
1 2	0 3	0 3 5	$2(0 \oplus 1 \oplus 3) \oplus 4(1 \vee 2)$
1 3	0 2	0 4 5	$\bar{1} \oplus 4 \oplus 5 \oplus 23$
1 4	$0\ \bar{5}$	1 2 3	$0(\bar{2} \oplus 3)$
1 5	0 4	1 2 4	$0(3 \oplus \bar{5})$
2 3	$0(2 \oplus 3)$	1 2 5	$0(3 \oplus 4)$
2 4	$0\ 3\ \bar{5}$	1 3 4	$0(2 \oplus \bar{5})$
2 5	0 3 4	1 3 5	$0(2 \oplus 4)$
3 4	$0\ 2\ \bar{5}$	1 4 5	$\bar{0} \oplus 4 \oplus 5$
3 5	0 2 4	2 3 4	$0(\bar{2} \oplus 3 \oplus 5)$
4 5	$\bar{5} \oplus 4$	2 3 5	$0(2 \oplus 3 \oplus 4)$
0 1 2	$0 \oplus 1 \oplus 3(\bar{0} \oplus 2)$	2 4 5	$03 \oplus 4 \oplus \bar{5}$
0 1 3	$0 \oplus 1 \oplus 2(\bar{0} \oplus 3)$	3 4 5	$02 \oplus 4 \oplus \bar{5}$
0 1 4	$0 \oplus 1 \oplus \bar{5} \oplus 23$		

Table 4.X.

$$f(\underline{a}) \neq g(\underline{a}) \quad .$$

A particular fault having a test function $t(\underline{x})=0$ for all $\underline{x}$ will be said *undetectable*. Another classical assumption of diagnosis theory that will be retained here is that all faults are detectable. In this case it will be possible to detect any fault in the network by the simple observation of the actual output $g(\underline{x})$ for any of the 2^n possible inputs and the comparison of these values with the prescribed values given by $f(\underline{x})$.

A set $\{\underline{a}\}$ of input vectors is called a test of a logic network if the observation of the corresponding outputs allows the detection of every possible fault in the network.

As the previous remark has shown, if all faults are detectable, the set of 2^n possible input vectors is a test of the network. However, if the network implementation (or scheme) is known, it will be possible to construct tests having less than 2^n elements. One of the objectives of the diagnosis theory is the construction of minimal tests : a test is called a *minimal test* if it ceases to be a test at the suppression of any single of its components. Search of minimal tests is usually carried out in two well-distinct parts :

(a) For each of the possible faults, research of the corresponding test vectors, i.e. computation of the corresponding test function.

(b) Research of a *minimal cover* of the faults by a set of test vectors. The term *minimal cover* has been used here to point out the similarity of this research with the more classical problem of covering minterms by prime implicants. In other terms, one searches the smallest possible number of test vectors which allows the detection of all faults within the network.

Various practical methods will be developed in the literature to solve this covering problem and we shall not deal with it any more, focusing our attention on the computation of test functions.

Consider two faults having for respective test functions $t_0(\underline{x})$ and $t_1(\underline{x})$. These faults are said to be *indistinguishable* if

$$t_0(\underline{x}) = t_1(\underline{x}) \ ;$$

they are distinguishable in the opposite case.

Indistinguishable **faults are for** instance the stuck-at-0 of either input of an AND-gate. Indistinguishability is an equivalence relation on the set of all possible faults : it is trivially reflexive, symmetric and transitive. The importance of indistinguishability comes from the following observation : the computation of the tests functions has only to be performed for a single representative of each of its equivalence classes.

4.4.3.2. Computation of test functions for simple faults.

Remember that a fault is called simple fault if only one (input) connection is stuck-at-0 or 1. It is assumed in fault detection theory that the simple faults occur the most frequently : it is easily understandable that in a network simple faults successively occur in the life of a circuit. Thus any multiple fault begins as being a simple fault and if the network is tested frequently enough this simple fault will be detected.

In fact the restriction of the diagnosis theory to the detection of simple faults is also due to the fact that simple faults constitute a small subset of the stuck-at-faults (there exists 2n single faults and 3^n-1 multiple faults) and that the test functions for the simple faults are easily computed.

Theorem 3.2.13. allows us to state that the test functions for the 2n simple faults are :

$$x_i \frac{\Delta f}{\Delta x_i}, \ \bar{x}_i \frac{\Delta f}{\Delta x_i}, \ i=0,1,\ldots,n-1.$$

Consider again the example 4.3.1.7 ; the differences computed in section 4.3.3.3. allow us to write the following test functions :

$$x_0 \frac{\Delta f}{\Delta x_0} = x_0(x_1 \oplus x_3x_2) \quad ; \quad \bar{x}_0 \frac{\Delta f}{\Delta x_0} = \bar{x}_0(x_1 \oplus x_3x_2)$$

$$x_1 \frac{\Delta f}{\Delta x_1} = x_1x_0 \quad ; \quad \bar{x}_1 \frac{f}{x_1} = \bar{x}_1x_0$$

$$x_2 \frac{\Delta f}{\Delta x_2} = x_2x_3x_0 \quad ; \quad \bar{x}_2 \frac{\Delta f}{\Delta x_2} = \bar{x}_2x_3x_0$$

$$x_3 \frac{\Delta f}{\Delta x_3} = x_3x_2x_0 \quad ; \quad \bar{x}_3 \frac{\Delta f}{\Delta x_3} = \bar{x}_3x_2x_0$$

$$x_4 \frac{\Delta f}{\Delta x_4} = x_4\bar{x}_5 \quad ; \quad \bar{x}_4 \frac{\Delta f}{\Delta x_4} = \bar{x}_4\bar{x}_5$$

$$x_5 \frac{\Delta f}{\Delta x_5} = x_5x_4 \quad ; \quad \bar{x}_5 \frac{\Delta f}{\Delta x_5} = \bar{x}_5x_4$$

We verify that the following set of test vectors t_i is a test for the set of simple faults :

$\{x_5x_4x_3x_2x_1x_0\} = \{(001101),(011011),(110101),(000010)\}$.

4.4.3.3. Computation of test functions for multiple faults

The computation of the test functions for multiple faults is most easily performed by starting from theorem 3.2.12 and 3.2.13. From theorem 3.2.13 we deduce indeed that the test functions are sensitivities while theorem 3.2.12.

gives us a formula for computing them.

Let $\underline{h}$ be a (fixed) value of $\underline{x}$; let us partition $\underline{x}$, and thus $\underline{h}$, into two subsets, i.e. : $\underline{x}=(\underline{x}_1,\underline{x}_0)$, $\underline{h}=(\underline{h}_1,\underline{h}_0)$. Theorems 3.2.12 and 3.2.13 allow us to write successively :

$$S^{\underline{h}}f/S\underline{x} = f(\underline{x}) \oplus f(\underline{h})$$

$$= \underset{\underline{e}_0}{\bar{\Phi}} \left(\frac{\Delta f}{\Delta \underline{x}_0^{\underline{e}_0}}\right)_{\underline{x}=\underline{h}} \left[\underset{x_i \in \underline{x}_0}{\wedge} (x_i \oplus h_i)^{e_i}\right] \oplus$$

$$\underset{\underline{e}_1}{\bar{\Phi}} \left(\frac{\Delta f}{\Delta \underline{x}_1^{\underline{e}_1}}\right)_{\underline{x}=\underline{h}} \left[\underset{x_i \in \underline{x}_1}{\wedge} (x_i \oplus h_i)^{e_i}\right] \oplus$$

$$\underset{\underline{e}_1,\underline{e}_0}{\bar{\Phi}} \left(\frac{\Delta f}{\Delta \underline{x}_1^{\underline{e}_1} \underline{x}_0^{\underline{e}_0}}\right)_{\underline{x}=\underline{h}} \left[\underset{x_i \in \underline{x}}{\wedge} (x_i \oplus h_i)^{e_i}\right] , \ \underline{e}_1,\underline{e}_0 \neq \underline{0}. \qquad (4.47)$$

$$f(\underline{h}_1,\underline{x}_0) \oplus f(\underline{h}) = \underset{\underline{e}_0}{\bar{\Phi}} \left(\frac{\Delta f}{\Delta \underline{x}_0^{\underline{e}_0}}\right)_{\underline{x}=\underline{h}} \left[\underset{x_i \in \underline{x}_0}{\wedge} (x_i \oplus h_i)^{e_i}\right] \qquad (4.48)$$

The modulo-2 sum of (4.47) and (4.48) gives us :

$$S^{\underline{h}_1}f/S\underline{x}_1 = f(\underline{x}) \oplus f(h_1,\underline{x}_0)$$

$$= \underset{\underline{e}_1}{\bar{\Phi}} \left(\frac{\Delta f}{\Delta \underline{x}_1^{\underline{e}_1}}\right)_{\underline{x}=\underline{h}} \left[\underset{x_i \in \underline{x}_1}{\wedge} (x_i \oplus h_i)^{e_i}\right] \oplus$$

$$\underset{e_1,e_0}{\bar{\Phi}} \left(\frac{\Delta f}{\Delta \underline{x}_1^{\underline{e}_1} \underline{x}_0^{\underline{e}_0}}\right)_{\underline{x}=\underline{h}} \left[\underset{x_i \in \underline{x}}{\wedge} (x_i \oplus h_i)^{e_i}\right] \qquad (4.49)$$

From (4.47) and (4.49) we deduce that the faults characterized by :

$$\underline{x}_1 \text{ is stuck-at } \underline{h}_1 \ \forall \ \underline{x}_1 \subseteq \underline{x}, \ \underline{h}_1 \subseteq \underline{h} \qquad (4.50)$$

have test functions that may immediately be deduced from the Newton expansion of f at $\underline{h}$. More precisely from (4.49) we deduce that $S^{\underline{h}_1}f/S\underline{x}_1$ is the modulo-2 sum of the terms of the Newton expansion of f at $\underline{h}$ from which we deduced the cubes formed by only letters of $\underline{x}_0$ (remember that $\underline{x}_0$ is the complementary subset of $\underline{x}_1$ with respect to x, i.e. : $\underline{x}_0=\underline{x}\setminus\underline{x}_1$).

Let us denote by $N_{\underline{h}}f$ the Newton expansion of f at $\underline{h}$; the following theorem derives from (4.47-4.49).

Theorem

The test functions for the 2^n-1 stuck faults : $\underline{x}_1$ stuck-at $\underline{h}_1$, $\underline{x}_1 \subseteq \underline{x} \neq \emptyset$ (there are 2^n-1 sets $\underline{x}_1$) are obtained from $N_{\underline{h}}f$ by deleting the constant term and the cubes formed by letters of $\underline{x}_0=\underline{x}\setminus\underline{x}_1$ only.

Let us illustrate the above theorem by means of a short example; consider the following Newton expansion of the three-variable Boolean function 3.1.14 at $x_i=0$:

$$N_{(000)}f = x_0x_2 \oplus x_1 \oplus x_1x_2 \quad . \tag{4.50}$$

Besides the function $N_{(000)}f$ which is the test function for x_i stuck-at 0 $\forall i$, one deduces from (4.50) the three following functions which are the test functions for the faults : a subset of (x_0,x_1,x_2) is stuck at 0 :

$$\{x_0x_2 \oplus x_1x_2, x_1 \oplus x_1x_2, x_0x_2\} \; .$$

The following algorithm, which derives from the theorem above, may then be used for obtaining the test functions for the 3^n-1 stuck faults at the inputs of an n-variable Boolean function $f(\underline{x})$.

4.4.3.4. Algorithm and examples

For each partition $(\underline{x}_0,\underline{x}_1)$ of $\underline{x}$ compute the Newton expansion $N_{\underline{h}_0\bar{\underline{h}}_1}f$ with

$\underline{h}_0=(h_0,h_1,\ldots,h_{p-1})$, $\bar{\underline{h}}_1=(\bar{h}_p,\bar{h}_{p+1},\ldots,\bar{h}_{n-1})$.

For each subset $\underline{x}_{00}$ of $\underline{x}_0$ compute the test functions obtained by deleting in $N_{\underline{h}_0\bar{\underline{h}}_1}f$ the cubes formed by letters of $\underline{x}_{00}$ only.

Let us briefly indicate a method of proof for this algorithm.

(i) For $\underline{x}_0=\underline{x}$, the Newton expansion is $N_{\underline{h}}f$; by putting $\underline{x}_1=\underline{x}\setminus\underline{x}_{00}$ and $\underline{x}_{00}=\underline{x}_0$ the statement in the algorithm reduces to that of the theorem above.

(ii) Step (i) detects all faults including only stuck faults at $x_i=h_i$.

Consider now the partitions of the form $(\underline{x}\setminus x_i, x_i)$; one then easily verifies (by using similar arguments as for the theorem) that the Newton expansions $N_{(\underline{h}/h_i,\bar{h}_i)}f$ allow us to detect any stuck fault including one fault $x_i=\bar{h}_i$. The proof of the algorithm derives then from an induction on the size of the set $\underline{x}_1=x_i$.

The above algorithm may finally be improved by observing that the test functions for $\underline{x}$ stuck-at $\underline{h}$ ($\forall$ $\underline{h}$ $\in$ the n-cube) are f and $\bar{f}$; this derives from the fact that the constant terms in the Newton expansions are either 0 or 1.

The above example will now be treated completely by using the algorithm.

- Let $\underline{h}=(000)$; $N_{000}f = x_0x_2 \oplus x_1 \oplus x_1x_2$;

  the test functions derived from $N_{000}f$ are obtained by deleting the cubes formed by subsets of $\underline{x}_{00}$ $\forall$ $\underline{x}_{00} \subseteq \underline{x}$, i.e.

$\{x_0x_2 \oplus x_1 \oplus x_1x_2 = f, x_0x_2 \oplus x_1x_2, x_1 \oplus x_1x_2, x_0x_2\}$;
the faults detected by these test functions are $x_0x_1x_2 = 000, x_0x_1 = 00$
$x_0x_2=00, x_1x_2 = 00, x_0 = 0, x_1 = 0, x_2 = 0$.

- $N_{100}f = x_2 \oplus x_2\bar{x}_0 \oplus x_1 \oplus x_1x_2$;
  test functions : $\{f, x_2 \oplus x_2\bar{x}_0 \oplus x_1x_2, x_2\bar{x}_0 \oplus x_1 \oplus x_1x_2, x_2\bar{x}_0\}$;
  faults detected : $x_0x_1x_2 = 100, x_0x_1 = 10, x_0x_2 = 10, x_0 = 1$.
- $N_{010}f = 1 \oplus x_2 \oplus x_2x_0 \oplus \bar{x}_1 \oplus \bar{x}_1x_2$;
  test functions :

  $$\{\bar{x}_1 \oplus x_2 \oplus x_2x_0 \oplus \bar{x}_1x_2 = \bar{f}, x_2x_0 \oplus \bar{x}_1 \oplus \bar{x}_1x_2, \bar{x}_1 \oplus \bar{x}_1x_2\};$$

  faults detected : $x_0x_1x_2 = 010, x_0x_1 = 01, x_1x_2 = 10, x_1 = 1$.
- $N_{001}f = x_0 \oplus \bar{x}_2x_0 \oplus \bar{x}_2x_1$;
  test functions : $\{f, \bar{x}_2x_0 \oplus \bar{x}_2x_1\}$;
  faultsdetected : $x_0x_1x_2 = 001, x_0x_2 = 01, x_1x_2 = 01, x_2 = 1$.
- $N_{110}f = 1 \oplus x_2\bar{x}_0 \oplus \bar{x}_1 \oplus \bar{x}_1x_2$;
  test function : $\bar{f}$;
  faults detected : $x_0x_1x_2 = 110, x_0x_1 = 11$.
- $N_{101}f = 1 \oplus \bar{x}_0 \oplus \bar{x}_0\bar{x}_2 \oplus \bar{x}_2 \oplus \bar{x}_2x_1$;
  test function : $\bar{f}$;
  faults detected : $x_0x_1x_2 = 101, x_0x_2 = 11$.
- $N_{011}f = \bar{x}_2 \oplus x_0 \oplus x_0\bar{x}_2 \oplus \bar{x}_1\bar{x}_2$;
  test functions : $\{f, \bar{x}_2 \oplus x_0\bar{x}_2 \oplus \bar{x}_1\bar{x}_2\}$;
  faults detected : $x_0x_1x_2 = 011, x_1x_2 = 11$.
- The fault $x_0x_1x_2 = 111$ is detected by the test function $f \oplus f(111)=\bar{f}$.

The above algorithm gives us 11 different test functions which are necessary for detecting the $3^3-1 = 26$ stuck-at faults ; these test functions are
$\{x_0x_2 \oplus x_1\bar{x}_2, x_0x_2 \oplus x_1x_2, x_1\bar{x}_2, x_0x_2, \bar{x}_0x_2 \oplus x_1\bar{x}_2, \bar{x}_0x_2, \bar{x}_0x_2 \oplus \bar{x}_1\bar{x}_2, \bar{x}_0x_2 \oplus \bar{x}_1x_2, \bar{x}_1\bar{x}_2,$
$x_0\bar{x}_2 \oplus x_1\bar{x}_2, \bar{x}_0\bar{x}_2 \oplus \bar{x}_1\bar{x}_2\}$.
One then verifies that the following set of test vectors is sufficient for detecting all stuck faults or, stated otherwise, that the test functions are equal to 1 for at least one of the test vectors :

$$\{x_0x_1x_2\} = \{010, 101, 110, 001\}.$$

Observe that this algorithm allows us to make automatically obvious the different equivalence classes of faults. Consider e.g. two subsets $\underline{x}_{01}$ and $\underline{x}_{02}$ of letters of $\underline{x}$; if the sets of cubes formed by letters of $\underline{x}_{01}$ and the sets of cubes formed by letters of $\underline{x}_{02}$ are identical in the Newton expansion of f at $\underline{h}$, the faults $f(\underline{h}_{01})$ and $f(\underline{h}_{02})$ are in the same equivalence class of faults.

$\underline{x}_0$	$\overset{0}{S} f/S\underline{x}_1 = f(\underline{x}_1,\underline{x}_0) \oplus f(\underline{0},\underline{x}_0)$	$\underline{x}_0$	$\overset{0}{S} f/S\underline{x}_1 = f(\underline{x}_1,\underline{x}_0) \oplus f(\underline{0},\underline{x}_0)$
∅	$10 \oplus 4 \oplus 54 \oplus 320$	2 3	$10 \oplus 4 \oplus 54 \oplus 320$
0	$10 \oplus 4 \oplus 54 \oplus 320$	2 4	$10 \oplus 54 \oplus 320$
1	$10 \oplus 4 \oplus 54 \oplus 320$	2 5	$10 \oplus 4 \oplus 54 \oplus 320$
2	$10 \oplus 4 \oplus 54 \oplus 320$	3 4	$10 \oplus 54 \oplus 320$
3	$10 \oplus 4 \oplus 54 \oplus 320$	3 5	$10 \oplus 4 \oplus 54 \oplus 320$
4	$10 \oplus 54 \oplus 320$	4 5	$10 \oplus 320$
5	$10 \oplus 4 \oplus 54 \oplus 320$	0 1 2	$4 \oplus 54 \oplus 320$
0 1	$4 \oplus 54 \oplus 320$	0 1 3	$4 \oplus 54 \oplus 320$
0 2	$10 \oplus 4 \oplus 54 \oplus 320$	0 1 4	$54 \oplus 320$
0 3	$10 \oplus 4 \oplus 54 \oplus 320$	0 1 5	$4 \oplus 54 \oplus 320$
0 4	$10 \oplus 54 \oplus 320$	0 2 3	$10 \oplus 4 \oplus 54$
0 5	$10 \oplus 4 \oplus 54 \oplus 320$	0 2 4	$10 \oplus 54 \oplus 320$
1 2	$10 \oplus 4 \oplus 54 \oplus 320$	0 2 5	$10 \oplus 54 \oplus 320$
1 3	$10 \oplus 4 \oplus 54 \oplus 320$	0 3 4	$10 \oplus 54 \oplus 320$
1 4	$10 \oplus 54 \oplus 320$	0 3 5	$10 \oplus 4 \oplus 54 \oplus 320$
1 5	$10 \oplus 4 \oplus 54 \oplus 320$	0 4 5	$10 \oplus 320$

Table 4.XI.

$\underline{x}_0$	$\overset{0}{S^{-}}f/S\underline{x}_1 = f(\underline{x}_1,\underline{x}_0) \oplus f(\underline{0},\underline{x}_0)$	$\underline{x}_0$	$\overset{0}{S^{-}}f/S\underline{x}_1 = f(\underline{x}_1,\underline{x}_0) \oplus f(\underline{0},\underline{x}_0)$
1 2 3	10 ⊕ 4 ⊕ 54 ⊕ 320	0 2 3 4	10 ⊕ 54
1 2 4	10 ⊕ 54 ⊕ 320	0 2 3 5	10 ⊕ 4 ⊕ 54
1 2 5	10 ⊕ 4 ⊕ 54 ⊕ 320	0 2 4 5	10 ⊕ 320
1 3 4	10 ⊕ 4 ⊕ 54 ⊕ 320	0 3 4 5	10 ⊕ 320
1 3 5	10 ⊕ 4 ⊕ 54 ⊕ 320	1 2 3 4	10 ⊕ 54 ⊕ 320
1 4 5	10 ⊕ 320	1 2 3 5	10 ⊕ 4 ⊕ 54 ⊕ 320
2 3 4	10 ⊕ 54 ⊕ 320	1 2 4 5	10 ⊕ 320
2 3 5	10 ⊕ 4 ⊕ 54 ⊕ 320	1 3 4 5	10 ⊕ 320
2 4 5	10 ⊕ 320	2 3 4 5	10 ⊕ 320
3 4 5	10 ⊕ 320	0 1 2 3 4	54
0 1 2 3	4 ⊕ 54	0 1 2 3 5	5 ⊕ 54
0 1 2 4	54 ⊕ 320	0 1 2 4 5	320
0 1 2 5	4 ⊕ 54 ⊕ 320	0 1 3 4 5	320
0 1 3 4	54 ⊕ 320	0 2 3 4 5	10
0 1 3 5	4 ⊕ 54 ⊕ 320	1 2 3 4 5	10
0 1 4 5	320	0 1 2 3 4 5	∅

Table 4.XI. (Continued)

4.4.3.5. Example 4.2.1.7. (Continued)

From the expansion of f at $\underline{0}$ we deduce the set of test functions for the faults : a subset of $\underline{x}$ is stuck-at $\underline{0}$. These test functions are gathered in Table 4.XI; these (2^6-1) faults are gathered into 8 equivalence classes each of which being characterized by a test function, i.e. :

$$x_1x_0 \oplus x_4 \oplus x_5x_4 \oplus x_3x_2x_0$$
$$x_1x_0 \oplus x_5x_4 \oplus x_3x_2x_0$$
$$x_1x_0 \oplus x_3x_2x_0$$
$$x_4 \oplus x_5x_4$$
$$x_5x_4 \oplus x_3x_2x_0$$
$$x_5x_4$$
$$x_3x_2x_0$$
$$x_1x_0$$

We verify that the following set of test vectors is a test for the faults under consideration.

$$\{(x_5x_4x_3x_2x_1x_0)\} = \{(111111), (011101), (011111)\} .$$

4.5. Detection of functional properties

4.5.1. Detection of the decomposition

4.5.1.1. Problem statement

The general problem of functional decomposition of Boolean functions has proven to be of high computational complexity. An approach to this problem is the decomposition chart of Ashenurst [1957]. Ashenurst's theory has been used by Curtis [1962] as a basis for a more general theory in order to develop a systematic method of deriving economical mutliple stage switching circuits. Unfortunately, the procedure proposed by Ashenurst and Curtis requires the testing of 2^n-2-n decomposition charts, where n is the number of input variables. Another approach has been proposed by Akers [1959] . The algebraic decomposition condition obtained by Akers however requires the computation of δ-δ-operators : this leads to algebraic expressions of overwhelming size.

In this section we show how an algebraisation of the Ashenurst's theory directly leads to simple decomposition conditions.

The algebraisation is performed through a systematic use of the Boolean operators. In particular, we present an algorithm for the simple, disjunctive or not, decompositions of Boolean functions when these are represented by the sets of all their prime implicants and of all their prime implicates.

4.5.1.2. Decomposition classification

The functional decomposition of Boolean functions is generally presented as a mathematical tool for designing economical combinational switching networks. Indeed, it is often possible to express a function $f(\underline{x})$ of n variables as a composite function of functions, as in the following equation for example :

$$f(\underline{x}) = F[y(\underline{x}_0), \underline{x}_1] \tag{4.51}$$

where $\underline{x}_0$ and $\underline{x}_1$ are subsets of the set of variables $\underline{x}$. Sometimes a composite expression can be found for a Boolean function f so that, in the composite expression, F and y are essentially simpler functions. Thus, if we wish to design a switching network for a Boolean function, we may accomplish this by designing networks for the several simpler functions of the composite representation.

Decompositions can be classified in different ways. According to the number of subfunctions, there are two types of decompositions : the *simple decomposition* and the *complex decomposition*. A simple decomposition is necessarily of the type (4.51), that is, F contains only one subfunction y. If F contains more than one subfunction the decomposition is said to be complex. The following definitions are also classical (Curtis [1962]) :

A *multiple decomposition* is characterized by an expression of the form :

$$f(\underline{x}) = F[y_0(\underline{x}_0),\ldots,y_m(\underline{x}_m),\underline{x}_{m+1}] \; , \; x_i \in \underline{x} \; . \tag{4.52}$$

An *iterative decomposition* is characterized by an expression of the form :

$$f(\underline{x}) = F[y_0(y_1(\ldots y_m(\underline{x}_m),\ldots \underline{x}_1),\underline{x}_0), \underline{x}_{m+1}] \, . \tag{4.53}$$

Multiple and iterative decompositions are evidently subclasses of the complex decomposition class. The decompositions can also be classified according to the fact that the sets $\underline{x}_i$ of variables are disjoint or not. If $\underline{x}_i \cap \underline{x}_j = \emptyset$ (empty set) for each pair $\underline{x}_i, \underline{x}_j \subseteq \underline{x}$ with $i \neq j$, the decomposition is said to be *disjunctive*. If not it is said to be *nondisjunctive*. The proper subset $\underline{x}_{ip}$ of $\underline{x}_i$ is classically defined as : $\underline{x}_{ip} = \underline{x}_i \cap (\cup \, \bar{\underline{x}}_j)$. The decompositions can finally be classified according to the fact that some of the proper subsets $\underline{x}_{ip}$ are empty or not. A decomposition is said to be *proper* if each of its proper subsets $\underline{x}_{ip}$ is non-empty. If not it is said to be *improper*. The simplest nontrivial improper decompositions of f are characterized by expressions of the form :

$$\begin{aligned} F(\underline{x}) &= F[y_0(\underline{x}_0), y_1(\underline{x}_0), \underline{x}_1] \; , \\ f(\underline{x}) &= F[y_0(y_1(\underline{x}_0), \underline{x}_1), \underline{x}_0] \; . \end{aligned} \tag{4.54}$$

According to the three ways of classification quoted above, a set of decomposition types can be defined as summarized in Table 4.XII. The way suggested in this section to solve the various kinds of decomposition problems arising by considering the different types of decomposition is noted in the entries of Table 4.XII. This will be detailed in the next sections. The following observations may however already be made. It is clear that disjunctive improper decompositions cannot exist. Moreover the simplest improper decompositions are all trivial ; they are necessarily of one of the three following forms :

$$\begin{aligned} f(\underline{x}) &= F[y(\underline{x}_0),\underline{x}] \\ f(\underline{x}) &= F[y(\underline{x}),\underline{x}_0] \qquad , x_0 \in \underline{x} \\ f(\underline{x}) &= F[y(\underline{x}),\underline{x}] \end{aligned} \tag{4.55}$$

	Disjunctive – Simple	Disjunctive – Complex	Nondisjunctive – Complex	Nondisjunctive – Simple
Improper	×	×		Trivial Decompositions
Proper	Basic Algorithm	Basic Algorithm and Theorems by Ashenurst	Basic Algorithm and Theorems by Curtis	Basic Algorithm

Table 4.XII. Types of decompositions

4.5.1.3. The results of Ashenurst and Curtis

Consider a simple decomposition of $f(\underline{x})$ which is characterized by a relation of the form (4.51). The set $\underline{x}_0$ will be called the bound set and the set $\underline{x}_1$ the free set. Ashenurst gave necessary and sufficient conditions for the existence of a simple disjunctive decomposition with a given bound set in terms of a decomposition chart.

Theorem

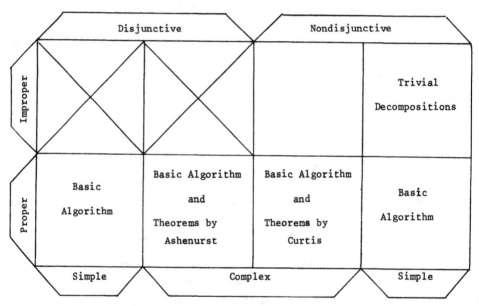

Let $\underline{x}_0,\underline{x}_1$ be a partition of $\underline{x}$. A switching function f is decomposable with bound set $\underline{x}_0$ and free set $\underline{x}_1$ if and only if the decomposition chart, with the variables in $\underline{x}_0$ defining the columns and the variables in $\underline{x}_1$ defining the rows, has at most four distinct kinds of rows which can be classified into the following categories :

1. All 0's.
2. All 1's.
3. A fixed pattern of 0's and 1's.
4. The complement of 3.

Ashenurst has stated a set of theorems relating simple disjunctive decompositions to complex ones. As a result it has been shown that we can derive all complex disjunctive decompositions of a given function from the set of all its simple decompositions. Therefore it is important to build algorithms for finding all simple decompositions of a given function.

The fundamental connection between simple disjunctive and nondisjunctive decompositions can be revealed by means of the following theorem :

Theorem

Let $\underline{x}_0,\underline{x}_1,\underline{x}_2$ be a partition of $\underline{x}$. A switching function f is decomposable with bound set $(\underline{x}_0,\underline{x}_2)$ and free set $(\underline{x}_1,\underline{x}_2)$ if and only if the 2^p subfunctions (p being the dimension of $\underline{x}_2$) $f(\underline{x}_0,\underline{x}_1,\underline{e}_2)$, $0 \leq \underline{e}_2 < 2^p$, are each decomposable with bound set $\underline{x}_0$ and free set $\underline{x}_1$.

Curtis has stated a set of theorems relating simple nondisjunctive decompositions to complex ones. All these theorems are in most cases immediate generalizations of the corresponding theorems by Ashenurst.

All types of decomposition quoted in Table 4.XII. are covered by the above theorem, except the improper complex decomposition. Until now a general theory for improper decomposition is, unavailable.

4.5.1.4. Fundamental theorems for simple decompositions.

We first state some theorems for simple decompositions (not necessarily disjunctive). Then we propose an algorithm for simple disjunctive decompositions.

Theorem (Thayse [1972c]) *Let $(\underline{x}_2, \underline{x}_1, \underline{x}_0)$ be a partition of the set of variables. A switching function f has a simple decomposition of the form*

$$f(\underline{x}) = F[\underline{x}_2,\underline{x}_1, y_0(\underline{x}_2,\underline{x}_0)]$$

iff for every x_i in $\underline{x}_0$ there exists a switching function g_i *, depending on the variables in $\underline{x}_0$ and $\underline{x}_2$ only, such that*

$$\frac{\Delta f}{\Delta x_i} = \frac{\delta f}{\delta \underline{x}_0} \; g_i(\underline{x}_2,\underline{x}_0) \; . \tag{4.56}$$

Proof.

The condition is necessary. Suppose indeed that

$$f(\underline{x}) = F[\underline{x}_2,\underline{x}_1,y_0(\underline{x}_2,\underline{x}_0)] \ ;$$

then the following two relations evidently hold :

$$\frac{\Delta f}{\Delta x_i} = \frac{\Delta F}{\Delta y_0}\ \frac{\Delta y_0}{\Delta x_i}\ , \quad \forall x_i \in \underline{x}_0$$

and

$$\frac{\delta f}{\delta \underline{x}_0} = \frac{\Delta F}{\Delta y_0}\frac{\delta y_0}{\delta \underline{x}_0} \ .$$

Since $(\delta y_0/\delta\underline{x}_0) \geqslant (\Delta y_0/\Delta x_i)$, for every x_i in $\underline{x}_0$, one has also

$$\frac{\Delta F}{\Delta y_0}\ \frac{\Delta y_0}{\Delta x_i} = \frac{\Delta F}{\Delta y_0}\ \frac{\Delta y_0}{\Delta x_i}\ \frac{\delta y_0}{\delta \underline{x}_0} = \frac{\delta f}{\delta \underline{x}_0}\frac{\Delta y_0}{\Delta x_i}\ ,$$

and thus

$$\frac{\Delta f}{\Delta x_i} = \frac{\delta f}{\delta \underline{x}_0}\ g_i(\underline{x}_2,\underline{x}_0) \tag{4.57}$$

where

$$g_i(\underline{x}_2,\underline{x}_0) = \frac{\Delta y_0}{\Delta x_i} \ .$$

The condition is sufficient. Suppose that condition (4.56) holds. From the fact that $\delta f/\delta \underline{x}_0$ of f with respect to $\underline{x}_0$ is a function that does not depend on $\underline{x}_0$, one deduces that

$$\frac{\Delta f}{\Delta \underline{x}_0^{\underline{e}_0}} = \frac{\delta f}{\delta \underline{x}_0}\ g_{\underline{e}_0}(\underline{x}_2,\underline{x}_0)$$

for every non-zero $\underline{e}_0$.

Then, from the partial Newton expansion of f with respect to $\underline{x}_0$ one deduces that

$$f(\underline{x}) = f(\underline{x}_2,\underline{x}_1,\underline{0}) \oplus \frac{\delta f}{\delta \underline{x}_0}\ y_0(\underline{x}_2,\underline{x}_0)$$

where

$$y_0(\underline{x}_2,\underline{x}_0) = \bigoplus_{\underline{e}_0 \in \underline{0}} g_{\underline{e}_0}(\underline{x}_2,\underline{0})\ \underline{x}_0^{\underline{e}_0} \ . \qquad \square$$

Condition (4.56) leads to several other decomposition conditions which were derived by different authors.

__Theorem.__ (Akers [1959]). *A switching function f has a simple decomposition of the form*

$$f(\underline{x}) = F[\underline{x}_2, \underline{x}_1, y_0(\underline{x}_2, \underline{x}_0)]$$

iff

$$\frac{\Delta f}{\Delta x_i} = \frac{\delta f}{\delta \underline{x}_0} \frac{\Delta y_0}{\Delta x_i}, \quad \forall\, x_i \in \underline{x}_0 \quad . \tag{4.58}$$

Proof. The proof is implied by the proof of the preceding theorem. □

This last condition is however of little practical use since it is generally assumed that the function y_0 is unknown ; therefore, Akers transformed the condition (4.58) into the following.

Theorem. (Akers [1959]). *A switching function f has a simple decomposition of the form*

$$f(\underline{x}) = F[\underline{x}_2, \underline{x}_1, y_0(\underline{x}_2, \underline{x}_0)]$$

iff

$$\frac{\delta}{\delta \underline{x}_1}\left(\frac{\delta f}{\delta \underline{x}_0}\right) \frac{\Delta f}{\Delta x_i} = \frac{\delta f}{\delta \underline{x}_0} \frac{\delta}{\delta \underline{x}_1}\left(\frac{\Delta f}{\Delta x_i}\right) \tag{4.59}$$

for every x_i *in* $\underline{x}_0$.

Proof. We only prove the necessity of that condition. The second part of the proof may be found in the original paper of Akers.

By making the operator $\delta/\delta\underline{x}_0$ act on both members of (4.58) we obtain :

$$\frac{\delta}{\delta \underline{x}_1}\left(\frac{\Delta f}{\Delta x_i}\right) = \frac{\delta}{\delta \underline{x}_1}\left(\frac{\delta f}{\delta \underline{x}_0}\right) \frac{\Delta y_0}{\Delta x_i} \quad ;$$

then, by multiplying both members of the last expression by $\delta f/\delta\underline{x}_0$ and by using again (4.58) we obtain the expected relation. □

Theorem. (Lapscher [1972]). *A switching function f has a simple decomposition of the form*

$$f(\underline{x}) = F[\underline{x}_2, \underline{x}_1, y_0(\underline{x}_2, \underline{x}_0)]$$

iff there exists a switching function A depending on the variables in $\underline{x}_2$ *and* $\underline{x}_1$ *only, and a switching function B depending on the variables in* $\underline{x}_2$ *and* $\underline{x}_0$ *only, such that*

$$f(\underline{x}_2, \underline{x}_1, \underline{x}_0) \oplus f(\underline{x}_2, \underline{x}_1, \underline{0}) = A(\underline{x}_2, \underline{x}_1) B(\underline{x}_2, \underline{x}_0) \quad .$$

Proof. The condition is obviously sufficient. It is necessary : the partial Newton expansion of F with respect to y_0 at point $y_0(\underline{x}_2, \underline{0})$ yields :

$$F(\underline{x}_2, \underline{x}_1, y_0) = F[\underline{x}_2, \underline{x}_1, y_0(\underline{x}_2, \underline{0})] \oplus [y_0 \oplus y_0(\underline{x}_2, \underline{0})] \frac{\Delta F}{\Delta y_0} \quad ;$$

it suffices to choose :

$$A(\underline{x}_2,\underline{x}_1) = \frac{\Delta F}{\Delta y_0}$$

and

$$B(\underline{x}_2,\underline{x}_0) = y_0(\underline{x}_2,\underline{x}_0) \oplus y_0(\underline{x}_2,\underline{0}) \ . \qquad \square$$

Note that a similar theorem was proved by Pichat [1968].

By taking the difference with respect to $x_k \in \underline{x}_1$ of both members of (4.56) we successively obtain :

$$\frac{\Delta f}{\Delta x_k x_i} = \frac{\Delta}{\Delta x_k} (\frac{\delta f}{\delta \underline{x}_0}) g_i$$

$$\frac{\Delta f}{\Delta x_j} \frac{\Delta f}{\Delta x_i x_k} = \frac{\delta f}{\delta \underline{x}_0} \frac{\Delta}{\Delta x_k} (\frac{\delta f}{\delta \underline{x}_0}) g_i g_j, \ (x_i, x_j \in \underline{x}_0, x_k \in \underline{x}_1)$$

and thus

$$\frac{\Delta f}{\Delta x_j} \frac{\Delta f}{\Delta x_i x_k} = \frac{\Delta f}{\Delta x_i} \frac{\Delta f}{\Delta x_j x_k}$$

This last relation constitutes the necessary conditions for disjunctive decompositions by Shen, McKellar and Weiner [1971] . The Akers and Shen theorems, initially proven only for the disjunctive case, remain valid thus even for the non-disjunctive one.

Algorithms for obtaining disjunctive decompositions of switching functions and which are based on the above theorems have been derived by Pichat [1968] , Shen et al. [1971], Thayse [1972c] and Lapscher [1972] . We now propose an algorithm grounded on (4.56) and on the computation formulae for obtaining the δ-functions (see section 3.2.7).

The use of (4.56) for the detection of simple decompositions requires that the set of δ-functions of a given switching function be easily available. We know that the knowledge of the prime implicants and of the prime implicates of a switching function allows an immediate computation of the variations in view of the formula (3.62), i.e. :

$$\frac{\delta f}{\delta \underline{x}_0} = (\frac{\overline{pf}}{p\underline{x}_0}) \frac{qf}{q\underline{x}_0} \ . \tag{4.60}$$

The formal expression of (3.62) will now be slightly modified in order to allow for building of iterative computation scheme for the set of variations of a switching function. Let us define $p_{\underline{x}_0}(f)$ and $q_{\underline{x}_0}(f)$ as follows : $p_{\underline{x}_0}(f)$ is the disjunction of all prime implicants of f degenerate in $\underline{x}_0$ and which contain all variables of $\underline{x}_1$ ($\underline{x}_1, \underline{x}_0$ is a partition of $\underline{x}$) ;

$q_{\underline{x}_0}(f)$ is the conjunction of all prime implicants of $\bar{f}$ degenerate in $\underline{x}_0$ and which contain all variables of $\underline{x}_1$ (see also theorem 4.3.1.3.).

The following equality then directly derives from (4.60) :

$$\frac{\delta f}{\delta \underline{x}_0} = \bigwedge_{x_i \in \underline{x}_1} \frac{\delta f}{\delta \underline{x}_0 x_i} [\overline{p_{\underline{x}_0}(f)}\overline{q_{\underline{x}_0}(f)}] . \qquad (4.61)$$

This expression allows us to compute immediately the functions $\delta f/\delta \underline{x}_0$ when the functions $\delta f/\delta \underline{x}_0 x_i$ and the prime implicates and implicants of f are known. Since for a non-degenerate switching function $\delta f/\delta \underline{x}=1$, all δ-functions may trivially be computed when starting with the δ-functions $\delta f/\delta \underline{x}_0$ with the highest cardinality of the set $\underline{x}_0$. Let us note that the computation scheme based on (4.61) requires to take into account each of the prime implicants and prime implicates once and only once.

4.5.1.5. Algorithm for disjunctive decomposition detection

(a) Compute the set of δ-functions of the switching function f to be tested by applying relation (4.61) if the sets of prime implicants and prime implicates of the function are known, by using the formula (3.60) otherwise. Since one is only concerned with non-trivial decompositions, that is decompositions whose bound set $\underline{x}_0$ contains at least two variables, only the δ-functions with respect to at most n-2 variables are to be computed.

(b) Verify if the conditions (4.56) are satisfied for the candidate bound sets $\underline{x}_0$. This verification can be performed, for example, by using the following computation scheme derived from (4.56). The δ-functions are naturally obtained by the intermediate of (4.61) as a product of implicates. One has :

$$\frac{\Delta f}{\Delta x_i} = \frac{\delta f}{\delta \underline{x}_0} y_i .$$

The prime implicants of $\delta f/\delta \underline{x}_0$ and of g_i are obtained from their conjunctive forms respectively. Each of the prime implicants of y_i is the product of two cubes $A_{ij}(\underline{x}_1)$ and $B_{ij}(\underline{x}_0)$, A_{ij} being the product of variables $x_j \in \underline{x}_1$ and $B_{ij}(\underline{x}_0)$ being the product variables $x_j \in \underline{x}_0$. The conditions (4.56) become then :

$$A_{ij}(\underline{x}_1) \geqslant \delta f/\delta \underline{x}_0 \quad , \qquad \forall\ i,j$$

or equivalently

$$\frac{\delta f}{\delta \underline{x}_0} [\bigvee_{i,j} \bar{A}_{ij}(\underline{x}_1)] = 0$$

which can be trivially verified.

4.5.1.6. Example 4.3.1.7. (Continued)

Consider the switching function of example 4.3.1.7. Obtain first this function as a disjunction of all its prime implicants and as a conjunction of all its prime implicates, i.e.

$$\begin{aligned} f = {} & \bar{x}_4\bar{x}_0 \vee \bar{x}_4\bar{x}_2\bar{x}_1 \vee \bar{x}_4x_3x_2x_1 \vee \bar{x}_4\bar{x}_3\bar{x}_1 \vee x_5\bar{x}_0 \vee x_5\bar{x}_2\bar{x}_1 \vee x_5x_3x_2x_1 \vee \\ & x_5\bar{x}_3\bar{x}_1 \vee \bar{x}_5x_4\bar{x}_2x_1x_0 \vee \bar{x}_5x_4x_3x_2\bar{x}_1x_0 \vee \bar{x}_5x_4\bar{x}_3x_1x_0 \\ = {} & (x_4 \vee \bar{x}_3 \vee \bar{x}_2 \vee x_1 \vee \bar{x}_0)(\bar{x}_5 \vee \bar{x}_3 \vee \bar{x}_2 \vee x_1 \vee \bar{x}_0)(x_5 \vee \bar{x}_4 \vee \bar{x}_3 \vee \bar{x}_2 \vee \bar{x}_1) \\ & (x_4 \vee x_2 \vee \bar{x}_1 \vee \bar{x}_0)(\bar{x}_5 \vee x_2 \vee \bar{x}_1 \vee \bar{x}_0)(x_4 \vee x_3 \vee \bar{x}_1 \vee \bar{x}_0) \\ & (\bar{x}_5 \vee x_3 \vee \bar{x}_1 \vee \bar{x}_0)(x_5 \vee \bar{x}_4 \vee x_2 \vee x_1)(x_5 \vee \bar{x}_4 \vee x_3 \vee x_1)(x_5 \vee \bar{x}_4 \vee x_0) \end{aligned}$$

The -functions are gathered in Table 4.7.

The following sets of variables satisfy the conditions (4.56)

$$(x_3,x_2,x_1,x_0),(x_3,x_2,x_1),(x_3,x_2),(x_5,x_4) \; .$$

The function f may, for example, be designed as follows :

$$\begin{aligned} f &= F\,[\,y(x_3,x_2,x_1,x_0),x_5,x_4] \\ F &= y \oplus (\bar{x}_4 \vee x_5) \\ y &= x_0(x_1 \oplus x_3x_2) \end{aligned}$$

4.5.2. Detection of symmetry

A function is partially symmetric with respect to two of its variables x_i and x_j if it does not change after a permutation of these variables; in order to express this invariance property, we shall write :

$$f(x_i,x_j) = f(x_j,x_i) \tag{4.62}$$

A function is partially symmetric in $\underline{x}_0$ ($\underline{x}_0 \subseteq \underline{x}$) if it is invariant for any permutation of the variables in $\underline{x}_0$.

In order to verify that a function is symmetric in $\underline{x}_0=(x_{p-1},x_{p-2}, \ldots, x_1,x_0)$ it is e.g. sufficient to verify that it is symmetric in (x_{p-1},x_{p-2}), $(x_{p-2},x_{p-3}),\ldots,(x_2,x_1)$ and (x_1,x_0). The detection of partial symmetry with respect to two variables may be grounded on the verification of the following formula. The partial Newton expansion with respect to the variables x_i,x_j and at the vertex x_j,x_i allows us to write :

$$f(x_i,x_j)=f(x_j,x_i)\oplus(x_i\oplus x_j)\frac{\Delta f}{\Delta x_i}\oplus(x_j\oplus x_i)\frac{\Delta f}{\Delta x_j}\oplus(x_i\oplus x_j)\frac{\Delta f}{\Delta x_i x_j}\quad ;$$

from the above equality we deduce :

$$f(x_i,x_j)\oplus f(x_j,x_i)=(x_i\oplus x_j)\frac{Sf}{Sx_ix_j} \tag{4.63}$$

and the function f is thus partially symmetric in x_i,x_j iff the following condition is satisfied :

$$(x_i\oplus x_j)\frac{Sf}{Sx_ix_j}=0 \tag{4.64}$$

We verify e.g. that for the example 4.3.1.7. :

$$(x_2\oplus x_3)\frac{Sf}{Sx_2x_3}=(x_2\oplus x_3)(x_3x_0\oplus x_2x_0\oplus x_0)\equiv 0.$$

Hence this function is partially symmetric in x_2,x_3.

For further information relative to symmetry detection see e.g. chapter 13 of the book by Davio, Deschamps and Thayse [1978] or the papers by Davio and Deschamps (Davio and Deschamps [1972], Deschamps [1973]). Let us simply mention that symmetry or partial symmetry of Boolean functions allows one to reduce the analysis and synthesis methods for these functions.

4.5.3. Detection of A-degeneracy

Let us first recall that the A-degeneracy gathers in a single definition the notions of degeneracy and of unateness. As for the other functional properties the detection of the A-degeneracy is of interest for the analysis and the synthesis of Boolean functions. This question has been extensively dealed with in chapter 12 of the book by Davio, Deschamps and Thayse [1978] . Let us only mention that the unateness property allows a simplification in the procedure of fault detection (Betancourt [1971]) and in the synthesis procedure ; unate functions are indeed realizable by means of monoform variables (i.e. variables in either direct form or complemented form) (see e.g. Mc Naughton [1961] , Ying and Susskind [1971]).

Theorem 3.2.16. provides us with three criteria allowing us to detect the A-degeneracy. Observe however that if we know either the set of prime implicants or the set of prime implicates of a Boolean function it is easier to detect directly the degeneracy and unateness without resorting to the evaluation of the criteria of theorem 3.2.16. Quine [1961] indeed showed that the representation of a monotone function as the disjunction of its prime implicants contains only variables either under direct form or under complemented form. The theorem by Quine allows us also to detect partial unateness which is generally a more common property than total unateness. A function is partially unate with respect to $\underline{x}_0 \subseteq \underline{x}$ if and only if this function is representable as a disjunction of implicants where the $x_i \in \underline{x}_0$ appear in only one form (Quine [1961]).

We verify e.g. that the function of example 4.3.1.7. is not unate.

4.5.4. Other applications

Let us mention that the applications presented in this chapter constitute only a small part of the Boolean calculus of differences. We have considered functional problems such as e.g. the detection of stuck-faults at the network inputs and the detection of function hazards.

We know that there exists similar problems related to the network realizing a Boolean function : these problems are classically referred to as logic problems. The detection of stuck-faults in the internal wiring of a network and the detection of logic hazards are the most usual logic problems. The solution of these problems by means of the Boolean calculus of differences may be found in several publications by the author.

We have only considered problems related to combinatorial networks. Several problems connected to sequential networks have been studied in the literature through the intermediate of the Boolean calculus of differences. Let us quote e.g. the detection of essential hazards (Unger [1959]), the detection of faults in sequential networks (Kohavi [1970]) and the transformation between several types of sequential networks (Thayse [1972b]).

Finally there exist several problems related to other fields than switching theory that could be treated by the intermediate of Boolean calculus of differences; let us quote e.g. fuzzy logic and multiple-valued algebra (Davio and Thayse [1973] , Symposia on multiple-valued logic [1975-1978]).

References

S. Akers,
1959, On a theory of Boolean functions, SIAM J., 7, 487-498.

V. Amar and N. Condulmari,
1967, Diagnosis of large combinational networks, IEEE Trans. Electron. Comput., EC-16, 675-680.

R. Ashenurst,
1957, The decomposition of switching functions, Paper presented at the international symposium on switching theory ; reprinted in appendix in : Curtis, A., A new approach to the design of switching circuits, (Van Nostrand, Princeton, 1962).

J. Beister,
1974, A unified approach to combinational hazards, IEEE Trans. Comput., C-21, 566-575.

R. Bellman,
1960, Introduction to matrix analysis, Mc Graw-Hill, New-York.

B. Benjauthrit and I. Reed,
1976, Galois switching functions and their applications, IEEE Trans. Comput., C-25, 78-86.

R. Betancourt,
1971, Derivation of minimum test sets for unate logical circuits, IEEE Trans. Comput., C-20, 1264-1269.

G. Bioul and M. Davio
1972, Taylor expansions of Boolean functions and of their derivatives, Philips Res. Rept., 27, 17-36.

G. Bioul, M. Davio and J.P. Deschamps,
1973, Minimization of ring-sum expansions of Boolean functions, Philips Res. Rept., 28, 17-36.

G. Birkhoff,
1967, Lattice theory, American mathematical society, Providence, USA.

D. Bochmann,
1971a, Dynamische operationen in der schaltalgebra, Nachrichtentechnik, 21,227-229.
1971b, Dynamische operationen in der schaltalgebra, Teil II, Nachrichtentechnik, 22, 282-283.

1972, Dynamische operationen in der schaltalgebra, Teil III, Nachrichtentechnik, 189-191.

1975a, Uber ableitungen in der schaltalgebra und einen damit formulierbaren entwicklungssutz, Wiss. Zeitschrift, TU Dresden, 14, N°5.

1975b, Einführung in die strukturelle automatentheorie, VEB Verlag Technik, Berlin, Hanser-Verlag, Munich and Vienna, 1975c, Binäre signale und systeme, Nachrichtentecnik-Elektronik, 25, N°5.

1978, Boolean differential calculus (A survey), Izv. Akad. Nauk SSSR Techn. Kibernet. 1977, N°5, 125-133 (Russian); translated as Engrg. Cybernet. 15, N°5, 67-75.

A. Brown and H.Young,

1970, Algebraic logic network analysis : towards an algebraic theory of the analysis and testing of digital networks, Report TR00-1974, IBM systems development division, Poughkeepsie, USA.

P. Calingaert,

1961, Switching functions : canonical forms based on commutative and associative binary operations, Trans. AIEE, 80, 808-814.

A. Curtis,

1962, A new approach to the design of switching circuits, Van Nostrand, Princeton.

M. Davio,

1961, Ring-sum expressions of Boolean functions, Proc. Symposium on computers and automata, 21, 411-418.

1973, Taylor expansions of symmetric Boolean functions, Philips Res. Rep., 28, 466-474.

M. Davio and G. Bioul,

1974, Taylor expansions computation from cube arrays, Philips Res. Rept., 29, 401-412.

M. Davio and J.P. Deschamps,

1972, Symmetric discrete functions, Philips Res. Rep., 27, 405-445.

M. Davio and Ph. Piret

1969, Les dérivées booléennes et leur application au diagnostic, Revue MBLE, 12, 63-76.

M. Davio and A. Thayse,

1973, Representation of fuzzy functions, Philips Res. Rept., 28, 93-106.

M. Davio, A. Thayse and G. Bioul,

1972, Symbolic computation of Fourier transforms of Boolean functions, Philips Res. Rept., 27, 386-403.

M. Davio, J. P. Deschamps and A.Thayse,
1978, Discrete and switching functions, Mc Graw-Hill, New York.

J. P. Deschamps,
1973, Partially symmetric switching functions, Philips Res. Rep., 28, 245-264.
1975, Binary simple decomposition of discrete functions, Digit. Proc., 1, 123-140.

J. P. Deschamps and A.Thayse,
1973a, On a theory of discrete functions, Part I : The lattice structure of discrete functions, Philips Res. Rept., 28, 397-423.
1973b, Applications of discrete functions, Part I :Transient analysis of combinational networks, Philips Res. Rept., 28, 497-529.
1975, Representation of discrete functions, International symposium on multiple-valued logic, 99-111.
1977, The module structure of discrete functions, International symposium on multiple-valued logic, 14-19.

E. Eichelberger,
1965, Hazard detection in combinational and sequential circuits, IBM J. Res. Develop., 9, 90-99.

A. Fadini,
1961, Operatori che estendono alle algebre di Boole la nozione di derivata, Giorn. mat. battaglini, 89, 42-64.

L. Fisher,
1974, Unateness properties of AND-EXCLUSIVE-OR logic circuits, IEEE Trans. Comput., C-23, 166-172.

M. Flomenhoft, S. Si, A. Susskind,
1973, Algebraic techniques for finding tests for several fault types, International symposium on fault-tolerant computing, 85-90.

H. Gazalé,
1959, Les structures de commutation à m valeurs et les calculatrices numériques, Gauthier-Villars, Paris.

G. Grätzer,
1971, Lattice theory, Freeman and company, San Francisco.

P. Hammer and S. Rudeanu,
1968, Boolean methods in operations research, Springer, Berlin.

M. Harrison

1965, Introduction to switching and automata theory, Mac Graw-Hill, New York.

M. Hsiao, F. Sellers and D. Chia,

1970, Fundamentals of Boolean difference for test pattern generation, Proc. 4th annual Princeton conference on information science and systems, 50-54.

D. Huffman,

1957, The design and use of hazard-free switching networks, J. of the ACM, 4, 47-62.

1958, Solvability criterion for simultaneous logical equations, MIT Research Lab. for electronics, quaterly progress report, 48-49.

C. Jordan,

1947, Calculus of finite differences, Chelsea publishing company, Chelsea.

A. Kaufmann,

1973, Théorie des sous-ensembles flous, Masson, Paris.

K. Kodandapani,

1976, Generalization of Reed-Muller canonical forms to multivalued logic, Thesis, University of Bangalore.

Z. Kohavi,

1970, Switching and finite automata theory, Mc Graw-Hill, New York.

C. Ku and G. Masson,

1975, The Boolean difference and multiple fault analysis, IEEE Trans. Comput., C-24, 62-71.

J. Kuntzmann,

1965, Algèbre de Boole, Dunod, Paris.

J. Kuntzmann and P. Naslin,

1967, Algèbre de Boole et machines logiques, Dunod, Paris.

F. Lapscher,

1972, Sur la recherche des décompositions disjointes d'une fonction booléenne, Rev. Fr. Automat. inf. rech. opér., 6, 92-112.

V. Lazarev and F. Piil,

1961, Synthesis method for finite automata, Automat. i telemeh., 22 1194-1201.

1962, A method for synthesizing finite automata with voltage-pulse feedback elements, Automat. i telemeh., 23, 1037-1043.

1963, The simplification of pulse-potential forms, Automat. i telemeh., 24, 271-276.

R. Lechner,
1971, Harmonic analysis of switching functions, in "Recent developments in switching theory", Ed. A. Mukhopadhyay, Academic Press, New York.

S. Lee
1978, Modern switching theory and digital design, Prentice-Hall, New Jersey.

D. Lewin,
1968, Logical design of switching circuits, Nelson, London.

D. Mange,
1973, Modèles asynchrones des bascules bistables, Systèmes logiques, Cahiers de la C.S.L., Numéro 5, 256-286.
1978, Analyse et synthèse des systèmes logiques, Editions Georgi, Lausanne.

M. Marcus and H. Minc,
1964, A survey of matrix theory and matrix inequalities, Allyn and Bacon, New York.

S. Marincovic and Z. Tosic,
1974, Algorithm for minimal polarized form determination, IEEE Trans. Comput., C-23, 1313-1315.

E. Mc Cluskey,
1965, Introduction to the theory of switching circuits, Mc Graw-Hill, New York.

R. Mc Naughton,
1961, Unate truth functions, IRE Trans. Electron. Comput., EC-10, 1-6.

A. Mukhopadhyay and G. Schmitz,
1970, Minimization of exclusive-or and logical equivalence switching circuits, IEEE Trans. Comput., C-19, 132-140.

D. Muller,
1954, Application of Boolean algebra to switching circuit design and to error detection, Trans. IRE, PGEC-3, 6-12.

E. Pichat,
1968, Décompositions simples de fonctions booléennes, Rev. Fr. Automat. inf. rech. opér., 2, 61-70.

D. Pradhan,
1978, A theory of Galois switching functions, IEEE Trans. Computers, C-27, 239-248.

F. Preparata and R. Yeh,
1972, On a theory of continuously valued logic, J. Comput. System Sci., 6, 397-418.
1973, Introduction to discrete structures, Addison-Wesley, Reading, USA.

W. Quine,
1955, A way to simplify truth functions, Am. math. Monthly, 62, 627-631.

S. Reddy,
1972, Easily testable realizations for logic functions, IEEE Trans. Comput., C-21, 1183-1188.

I. Reed,
1954, A class of multiple-error-correcting codes and the decoding scheme, Trans. IRE, PGIT 4, 38-49.

I. Rosenberg,
1974, Minimization of pseudo-Boolean functions by binary development, Discrete Math., 7, 151-165.

S. Rudeanu,
1961, On the definition of Boolean algebras by means of binary operations, Rev. Math. pures appl., 6, pp. 71-183.
1974, Boolean functions and equations, North-Holland, Amsterdam.

F. Sellers, M. Hsiao and L. Bearnson,
1968, Analysing errors with the Boolean difference, IEEE Trans. Comput., C-17, 676-683.

C. Shannon,
1938, A symbolic analysis of relay and switching circuits, Trans. of the AIEE, 57, 713-723.
1948, A mathematical theory of communication, Bell system Tech. J., 27, 279-423 and 623-656.

V. Shen, A. Mc Kellar and P. Weiner,
1971, A fast algorithm for the disjunctive decomposition of switching functions, IEEE Trans. Comput., C-20, 304-309.

R. Sikorski,
1969, Boolean algebras, Springer-Verlag, Berlin.

J. Smith and C. Roth,
1971, Analysis and synthesis of asynchronous sequential networks using edge-sensitive flip-flops, IEEE Trans. Comput., C-20, 847-855.

M. Stone,

1935, Subsumption of Boolean algebras under the theory of rings, Proc. Nat. Acad. Sci., USA, 21, 103-105.

A. Talantsev,

1959, On the analysis and synthesis of certain electrical circuits by means of special logical operators, Automat. i telemeh., 20, 898-907.

A. Thayse,

1970, Transient analysis of logical networks applied to hazard detection, Philips Res. Rep., 25, 261-336.

1971, Boolean differential calculus, Philips Res. Rep., 26, 229-246.

1972a, A variational diagnosis method for stuck-faults in combinatorial networks, Philips Res. Rep. 27, 82-98.

1972b, Testing of asynchronous sequential switching circuits, Philips Res. Rep., 27, 99-106.

1972c, A fast algorithm for the proper decomposition of Boolean functions, Philips Res. Rep., 27, 140-150.

1972d, Multiple-fault detection in large logical networks, Philips Res. Rep., 27, 583-602.

1973a, Disjunctive and conjunctive operators for Boolean functions, Philips Res. Rep., 28, 1-16.

1973b, On some iteration properties of Boolean functions, Philips Res. Rep., 28, 107-119.

1974a, New method for obtaining the optimal Taylor expansions of a Boolean function, Electron. Lett., 10, n°25/26, 543-544.

1974b, Differential calculus for functions from $(GF(p))^n$ to GF(p), Philips Res. Rep., 29, 560-586.

1974c, Applications of discrete functions, Part II, Transient analysis of asynchronous switching networks, Philips Res. Rep., 29, 155-192.

1974d, On a theory of discrete functions, Part IV, Discrete functions of functions, Philips Res. Rep., 29, 305-329.

1974e, Applications of discrete functions, Part III, The use of functions of functions in switching circuits, Philips Res. Rep., 29, 429-452.

1975a, La détection des aléas dans les circuits logiques au moyen du calcul différentiel booléen, Digit. Process., 1, 141-169.

1975b, Cellular hazard-free design of combinatorial networks, MBLE Internal Rep., R293.

1976, Difference operators and extended truth vectors for discrete functions, Discrete math., 14, 171-202.

1977, Static-Hazard detection in switching circuits by prime-implicant examination in fuzzy functions, Electronics letters, 13, n°4, 94-96.

1978, Meet and join derivatives and their use in switching theory, IEEE Trans. comput., C-27, 713-720.

1979a, Expansions of discrete functions in powers of an integer, Discrete applied mathematics, to appear.

1979b, Integer expansions of switching functions and their use in optimization problems, International symposium on multiple-valued logic.

A. Thayse, M. Davio,

1973, Boolean differential calculus and its application to switching theory, IEEE Trans. Computers, C-22, 409-420.

A. Thayse, M. Davio and J.P. Deschamps,

1978, Optimization of multivalued decision algorithms, International symposium on multiple-valued logic, 171-178.

A. Thayse and J.P. Deschamps,

1973, On a theory of discrete functions, Part II, The ring and field structures of discrete functions, Philips Res. Rep., 28, 424-465.

1976, Logic properties of unate and of symmetric discrete functions, International symposium on multiple-valued logic, 79-87.

1977, Logic properties of unate discrete and switching functions, IEEE Trans. Computers, C-26, 1202-1212.

P. Tison,

1965, Théorie des consensus, Dissertation doctorale, Grenoble.

1967, Generalization of consensus theory and application to the minimization of Boolean functions, IEEE Trans. Electron. Comput., EC-5, 126-132.

1971, An algebra for logic systems ; switching circuits application, IEEE Trans. Computers, C-20, 339-351.

P. Tosic,

1972, Analytical representations of m-valued logical functions over the ring of integers modulo m, Publications de la faculté d'électronique de l'université de Belgrade, n°410-425.

J. Tucker,

1974, A transition calculus for Boolean functions, Virginia polytechnic Microfilms Xerox univ., Ann. Harbor, USA.

S. Unger,

1959, Hazards and delays in asynchronous sequential switching circuits, Trans. Circuit Theory, CT-6, 12-25.

S. Yau and Y. Tang,
1971, An efficient algorithm for generating complete test sets for combinatorial logic circuit, IEEE Trans. Computers, C-20, 1245-1251.

C. Ying and A. Susskind,
1971, Building blocks and synthesis techniques for the realization of M-ary combinational switching functions, International symposium on multiple-valued logic, 141-149.

L. Zadeh,
1965, Fuzzy sets, Information and Control, 8, 338-353.

Author Index

Subject Index

List of symbols

Conjunction : No symbol or $\wedge$

Disjunction : $\vee$ or $\bigvee$

Negation of a : $\bar{a}$

Ring sum or addition modulo-2 : $\oplus$ or $\bigoplus$

Sum in a finite field : + or $\sum$

Matrix product with additive law A and multiplicative law M : No symbol (only if the additive and multiplicative laws are unambiguously defined by the context) or [AM] .

Kronecker matrix product with multiplicative law M : $\otimes$ (only if the multiplicative law is unambiguously defined by the context) or $\overset{M}{\otimes}$.

Union of sets : $\cup$

Intersection of sets : $\cap$

Inclusion of a set in another set : $\subseteq$ or $\subset$ (strict inclusion)

$\in$ means is an element of

Boolean constants are denoted by lower-case letters like a,...,e,...ℓ .

Boolean variables are denoted by lower-case letters like x,y,z.

Bold-face lower-case letters like $\underline{e}$ are used to denote a vector of discrete constants.

Bold-face lower-case letters like $\underline{x}$ are used to denote a vector of discrete variables.

Lattice exponentiation $x^{(a)}$: $x^{(a)} = 1$ iff $x \in a \quad \forall a \subseteq \{0,1\}$

$= 0$ otherwise.

Vol. 49: Interactive Systems. Proceedings 1976. Edited by A. Blaser and C. Hackl. VI, 380 pages. 1976.

Vol. 50: A. C. Hartmann, A Concurrent Pascal Compiler for Minicomputers. VI, 119 pages. 1977.

Vol. 51: B. S. Garbow, Matrix Eigensystem Routines – Eispack Guide Extension. VIII, 343 pages. 1977.

Vol. 52: Automata, Languages and Programming. Fourth Colloquium, University of Turku, July 1977. Edited by A. Salomaa and M. Steinby. X, 569 pages. 1977.

Vol. 53: Mathematical Foundations of Computer Science. Proceedings 1977. Edited by J. Gruska. XII, 608 pages. 1977.

Vol. 54: Design and Implementation of Programming Languages. Proceedings 1976. Edited by J. H. Williams and D. A. Fisher. X, 496 pages. 1977.

Vol. 55: A. Gerbier, Mes premières constructions de programmes. XII, 256 pages. 1977.

Vol. 56: Fundamentals of Computation Theory. Proceedings 1977. Edited by M. Karpiński. XII, 542 pages. 1977.

Vol. 57: Portability of Numerical Software. Proceedings 1976. Edited by W. Cowell. VIII, 539 pages. 1977.

Vol. 58: M. J. O'Donnell, Computing in Systems Described by Equations. XIV, 111 pages. 1977.

Vol. 59: E. Hill, Jr., A Comparative Study of Very Large Data Bases. X, 140 pages. 1978.

Vol. 60: Operating Systems, An Advanced Course. Edited by R. Bayer, R. M. Graham, and G. Seegmüller. X, 593 pages. 1978.

Vol. 61: The Vienna Development Method: The Meta-Language. Edited by D. Bjørner and C. B. Jones. XVIII, 382 pages. 1978.

Vol. 62: Automata, Languages and Programming. Proceedings 1978. Edited by G. Ausiello and C. Böhm. VIII, 508 pages. 1978.

Vol. 63: Natural Language Communication with Computers. Edited by Leonard Bolc. VI, 292 pages. 1978.

Vol. 64: Mathematical Foundations of Computer Science. Proceedings 1978. Edited by J. Winkowski. X, 551 pages. 1978.

Vol. 65: Information Systems Methodology, Proceedings, 1978. Edited by G. Bracchi and P. C. Lockemann. XII, 696 pages. 1978.

Vol. 66: N. D. Jones and S. S. Muchnick, TEMPO: A Unified Treatment of Binding Time and Parameter Passing Concepts in Programming Languages. IX, 118 pages. 1978.

Vol. 67: Theoretical Computer Science, 4th GI Conference, Aachen, March 1979. Edited by K. Weihrauch. VII, 324 pages. 1979.

Vol. 68: D. Harel, First-Order Dynamic Logic. X, 133 pages. 1979.

Vol. 69: Program Construction. International Summer School. Edited by F. L. Bauer and M. Broy. VII, 651 pages. 1979.

Vol. 70: Semantics of Concurrent Computation. Proceedings 1979. Edited by G. Kahn. VI, 368 pages. 1979.

Vol. 71: Automata, Languages and Programming. Proceedings 1979. Edited by H. A. Maurer. IX, 684 pages. 1979.

Vol. 72: Symbolic and Algebraic Computation. Proceedings 1979. Edited by E. W. Ng. XV, 557 pages. 1979.

Vol. 73: Graph-Grammars and Their Application to Computer Science and Biology. Proceedings 1978. Edited by V. Claus, H. Ehrig and G. Rozenberg. VII, 477 pages. 1979.

Vol. 74: Mathematical Foundations of Computer Science. Proceedings 1979. Edited by J. Bečvář. IX, 580 pages. 1979.

Vol. 75: Mathematical Studies of Information Processing. Proceedings 1978. Edited by E. K. Blum, M. Paul and S. Takasu. VIII, 629 pages. 1979.

Vol. 76: Codes for Boundary-Value Problems in Ordinary Differential Equations. Proceedings 1978. Edited by B. Childs et al. VIII, 388 pages. 1979.

Vol. 77: G. V. Bochmann, Architecture of Distributed Computer Systems. VIII, 238 pages. 1979.

Vol. 78: M. Gordon, R. Milner and C. Wadsworth, Edinburgh LCF. VIII, 159 pages. 1979.

Vol. 79: Language Design and Programming Methodology. Proceedings, 1979. Edited by J. Tobias. IX, 255 pages. 1980.

Vol. 80: Pictorial Information Systems. Edited by S. K. Chang and K. S. Fu. IX, 445 pages. 1980.

Vol. 81: Data Base Techniques for Pictorial Applications. Proceedings, 1979. Edited by A. Blaser. XI, 599 pages. 1980.

Vol. 82: J. G. Sanderson, A Relational Theory of Computing. VI, 147 pages. 1980.

Vol. 83: International Symposium Programming. Proceedings, 1980. Edited by B. Robinet. VII, 341 pages. 1980.

Vol. 84: Net Theory and Applications. Proceedings, 1979. Edited by W. Brauer. XIII, 537 Seiten. 1980.

Vol. 85: Automata, Languages and Programming. Proceedings, 1980. Edited by J. de Bakker and J. van Leeuwen. VIII, 671 pages. 1980.

Vol. 86: Abstract Software Specifications. Proceedings, 1979. Edited by D. Bjørner. XIII, 567 pages. 1980

Vol. 87: 5th Conference on Automated Deduction. Proceedings, 1980. Edited by W. Bibel and R. Kowalski. VII, 385 pages. 1980.

Vol. 88: Mathematical Foundations of Computer Science 1980. Proceedings, 1980. Edited by P. Dembiński. VIII, 723 pages. 1980.

Vol. 89: Computer Aided Design - Modelling, Systems Engineering, CAD-Systems. Proceedings, 1980. Edited by J. Encarnacao. XIV, 461 pages. 1980.

Vol. 90: D. M. Sandford, Using Sophisticated Models in Resolution Theorem Proving.
XI, 239 pages. 1980

Vol. 91: D. Wood, Grammar and L Forms: An Introduction. IX, 314 pages. 1980.

Vol. 92: R. Milner, A Calculus of Communication Systems. VI, 171 pages. 1980.

Vol. 93: A. Nijholt, Context-Free Grammars: Covers, Normal Forms, and Parsing. VII, 253 pages. 1980.

Vol. 94: Semantics-Directed Compiler Generation. Proceedings, 1980. Edited by N. D. Jones. V, 489 pages. 1980.

Vol. 95: Ch. D. Marlin, Coroutines. XII, 246 pages. 1980.

Vol. 96: J. L. Peterson, Computer Programs for Spelling Correction: VI, 213 pages. 1980.

Vol. 97: S. Osaki and T. Nishio, Reliability Evaluation of Some Fault-Tolerant Computer Architectures. VI, 129 pages. 1980.

Vol. 98: Towards a Formal Description of Ada. Edited by D. Bjørner and O. N. Oest. XIV, 630 pages. 1980.

Vol. 99: I. Guessarian, Algebraic Semantics. XI, 158 pages. 1981.

Vol. 100: Graphtheoretic Concepts in Computer Science. Edited by H. Noltemeier. X, 403 pages. 1981.

Vol. 101: A. Thayse, Boolean Calculus of Differences. VII, 144 pages. 1981.

This series reports new developments in computer science research and teaching – quickly, informally and at a high level. The type of material considered for publication includes:

1. Preliminary drafts of original papers and monographs
2. Lectures on a new field or presentations of a new angle in a classical field
3. Seminar work-outs
4. Reports of meetings, provided they are
   a) of exceptional interest and
   b) devoted to a single topic.

Texts which are out of print but still in demand may also be considered if they fall within these categories.

The timeliness of a manuscript is more important than its form, which may be unfinished or tentative. Thus, in some instances, proofs may be merely outlined and results presented which have been or will later be published elsewhere. If possible, a subject index should be included. Publication of Lecture Notes is intended as a service to the international computer science community, in that a commercial publisher, Springer-Verlag, can offer a wide distribution of documents which would otherwise have a restricted readership. Once published and copyrighted, they can be documented in the scientific literature.

Manuscripts

Manuscripts should be no less than 100 and preferably no more than 500 pages in length.
They are reproduced by a photographic process and therefore must be typed with extreme care. Symbols not on the typewriter should be inserted by hand in indelible black ink. Corrections to the typescript should be made by pasting in the new text or painting out errors with white correction fluid. Authors receive 75 free copies and are free to use the material in other publications. The typescript is reduced slightly in size during reproduction; best results will not be obtained unless the text on any one page is kept within the overall limit of 18 x 26.5 cm (7 x 10½ inches). On request, the publisher will supply special paper with the typing area outlined.
Manuscripts should be sent to Prof. G. Goos, Institut für Informatik, Universität Karlsruhe, Zirkel 2, 7500 Karlsruhe/Germany, Prof. J. Hartmanis, Cornell University, Dept. of Computer-Science, Ithaca, NY/USA 14850, or directly to Springer-Verlag Heidelberg.

Springer-Verlag, Heidelberger Platz 3, D-1000 Berlin 33
Springer-Verlag, Neuenheimer Landstraße 28–30, D-6900 Heidelberg 1
Springer-Verlag, 175 Fifth Avenue, New York, NY 10010/USA

ISBN 3-540-10286-8
ISBN 0-387-10286-8